T0203566

MONOGRAPHS ON STATISTICS AND APPLIED PROBABILITY

General Editors

V. Isham, N. Keiding, T. Louis, N. Reid, R. Tibshirani, and H. Tong

35 Empirical Bayes Methods, 2nd edition *J.S. Maritz and T. Lwin* (1989)

36 Symmetric Multivariate and Related Distributions *K.T. Fang, S. Kotz and K.W. Ng* (1990)

37 Generalized Linear Models, 2nd edition *P. McCullagh and J.A. Nelder* (1989)

38 Cyclic and Computer Generated Designs, 2nd edition *J.A. John and E.R. Williams* (1995)

39 Analog Estimation Methods in Econometrics *C.F. Manski* (1988)

40 Subset Selection in Regression *A.J. Miller* (1990)

41 Analysis of Repeated Measures *M.J. Crowder and D.J. Hand* (1990)

42 Statistical Reasoning with Imprecise Probabilities *P. Walley* (1991)

43 Generalized Additive Models *T.J. Hastie and R.J. Tibshirani* (1990)

44 Inspection Errors for Attributes in Quality Control *N.L. Johnson, S. Kotz and X. Wu* (1991)

45 The Analysis of Contingency Tables, 2nd edition *B.S. Everitt* (1992)

46 The Analysis of Quantal Response Data *B.J.T. Morgan* (1992)

47 Longitudinal Data with Serial Correlation—A state-space approach *R.H. Jones* (1993)

48 Differential Geometry and Statistics *M.K. Murray and J.W. Rice* (1993)

49 Markov Models and Optimization *M.H.A. Davis* (1993)

50 Networks and Chaos—Statistical and probabilistic aspects *O.E. Barndorff-Nielsen, J.L. Jensen and W.S. Kendall* (1993)

51 Number-Theoretic Methods in Statistics *K.-T. Fang and Y. Wang* (1994)

52 Inference and Asymptotics *O.E. Barndorff-Nielsen and D.R. Cox* (1994)

53 Practical Risk Theory for Actuaries *C.D. Daykin, T. Pentikäinen and M. Pesonen* (1994)

54 Biplots *J.C. Gower and D.J. Hand* (1996)

55 Predictive Inference—An introduction *S. Geisser* (1993)

56 Model-Free Curve Estimation *M.E. Tarter and M.D. Lock* (1993)

57 An Introduction to the Bootstrap *B. Efron and R.J. Tibshirani* (1993)

58 Nonparametric Regression and Generalized Linear Models *P.J. Green and B.W. Silverman* (1994)

59 Multidimensional Scaling *T.F. Cox and M.A.A. Cox* (1994)

60 Kernel Smoothing *M.P. Wand and M.C. Jones* (1995)

61 Statistics for Long Memory Processes *J. Beran* (1995)

62 Nonlinear Models for Repeated Measurement Data *M. Davidian and D.M. Giltinan* (1995)

63 Measurement Error in Nonlinear Models *R.J. Carroll, D. Rupert and L.A. Stefanski* (1995)

64 Analyzing and Modeling Rank Data *J.J. Marden* (1995)

65 Time Series Models—In econometrics, finance and other fields *D.R. Cox, D.V. Hinkley and O.E. Barndorff-Nielsen* (1996)

66 Local Polynomial Modeling and its Applications *J. Fan and I. Gijbels* (1996)

67 Multivariate Dependencies—Models, analysis and interpretation *D.R. Cox and N. Wermuth* (1996)

68 Statistical Inference—Based on the likelihood *A. Azzalini* (1996)
69 Bayes and Empirical Bayes Methods for Data Analysis
B.P. Carlin and T.A Louis (1996)
70 Hidden Markov and Other Models for Discrete-Valued Time Series
I.L. Macdonald and W. Zucchini (1997)
71 Statistical Evidence—A likelihood paradigm *R. Royall* (1997)
72 Analysis of Incomplete Multivariate Data *J.L. Schafer* (1997)
73 Multivariate Models and Dependence Concepts *H. Joe* (1997)
74 Theory of Sample Surveys *M.E. Thompson* (1997)
75 Retrial Queues *G. Falin and J.G.C. Templeton* (1997)
76 Theory of Dispersion Models *B. Jørgensen* (1997)
77 Mixed Poisson Processes *J. Grandell* (1997)
78 Variance Components Estimation—Mixed models, methodologies and applications
P.S.R.S. Rao (1997)
79 Bayesian Methods for Finite Population Sampling
G. Meeden and M. Ghosh (1997)
80 Stochastic Geometry—Likelihood and computation
O.E. Barndorff-Nielsen, W.S. Kendall and M.N.M. van Lieshout (1998)
81 Computer-Assisted Analysis of Mixtures and Applications—
Meta-analysis, disease mapping and others *D. Böhning* (1999)
82 Classification, 2nd edition *A.D. Gordon* (1999)
83 Semimartingales and their Statistical Inference *B.L.S. Prakasa Rao* (1999)
84 Statistical Aspects of BSE and vCJD—Models for Epidemics
C.A. Donnelly and N.M. Ferguson (1999)
85 Set-Indexed Martingales *G. Ivanoff and E. Merzbach* (2000)
86 The Theory of the Design of Experiments *D.R. Cox and N. Reid* (2000)
87 Complex Stochastic Systems
O.E. Barndorff-Nielsen, D.R. Cox and C. Klüppelberg (2001)
88 Multidimensional Scaling, 2nd edition *T.F. Cox and M.A.A. Cox* (2001)
89 Algebraic Statistics—Computational Commutative Algebra in Statistics
G. Pistone, E. Riccomagno and H.P. Wynn (2001)
90 Analysis of Time Series Structure—SSA and Related Techniques
N. Golyandina, V. Nekrutkin and A.A. Zhigljavsky (2001)
91 Subjective Probability Models for Lifetimes
Fabio Spizzichino (2001)
92 Empirical Likelihood *Art B. Owen (2001)*
93 Statistics in the 21st Century *Adrian E. Raftery, Martin A. Tanner, and Martin T. Wells (2001)*
94 Accelerated Life Models: Modeling and Statistical Analysis
Vilijandas Bagdonavičius and Mikhail Nikulin (2001)
95 Subset Selection in Regression, Second Edition
Alan Miller (2002)
96 Topics in Modelling of Clustered Data
Marc Aerts, Helena Geys, Geert Molenberghs, and Louise M. Ryan (2002)
97 Components of Variance *D.R. Cox and P.J. Solomon* (2002)

Longitudinal Data with Serial Correlation: A State-space Approach

R.H. JONES

Professor of Biometrics and
Director of the Biometrics Graduate Program
University of Colorado School of Medicine
Denver, USA

CRC Press
Taylor & Francis Group
Boca Raton London New York

CRC Press is an imprint of the
Taylor & Francis Group, an **informa** business
A CHAPMAN & HALL BOOK

First edition 1993
Originally published by Chapman & Hall

Published 2000 by CRC Press
Taylor & Francis Group
6000 Broken Sound Parkway NW, Suite 300
Boca Raton, FL 33487-2742

CRC Press is an imprint of Taylor & Francis Group, an Informa business

First issued in paperback 2019

ISBN 13: 978-0-367-45008-3 (pbk)
ISBN 13: 978-0-412-40650-8 (hbk)

Visit the Taylor & Francis Web site at
http://www.taylorandfrancis.com

and the CRC Press Web site at
http://www.crcpress.com

Library of Congress Cataloging-in-Publication Data

Catalog record is available from the Library of Congress.

TO JULIE

Contents

Preface

Longitudinal data analysis is the analysis of data collected on subjects who are followed over time. The data from many subjects are analyzed as a single analysis. Repeated measures analysis and growth curve analysis are special cases of longitudinal data analysis. Often, the main purpose of such an analysis is to determine whether groups of subjects respond differently to different treatments. Because there are multiple observations on each subject, these observations tend to be correlated, and this correlation must be accounted for in order to produce a proper analysis. In this monograph, the usual assumption is that the errors have Gaussian distributions, and the estimation approach is maximum likelihood. In all but the simplest situations, maximum likelihood estimation requires searching over the parameter space in an attempt to find the values of the unknown parameters that maximize the likelihood function. Since algorithms for nonlinear optimization are readily available, the difficult part often is to express the likelihood in a form that can be calculated. When the data from each subject are serially correlated, the state space approach provides a convenient way to compute likelihoods using the Kalman filter.

This monograph is written for students at the graduate level in biostatistics, statistics or other disciplines that collect longitudinal data. It assumes that the reader is familiar with statistical theory and methods at the first year graduate level, particularly the matrix approach to regression analysis. It is not assumed that the reader is familiar with state space methodology. The first two chapters present the background of longitudinal data analysis. Serially correlated within subject errors are introduced in Chapter 3. Finally, in Chapter 4, the state space representation and the Kalman filter are introduced. Chapter 5 brings together state space representations and longitudinal data models. Chapter 6 is a more advanced topic using more complicated within subject error models and can be skipped without affecting the readers ability to follow the last two chapters. Chapter 7 introduces models where the random effects appear nonlinearly. This is currently an area of intensive research, and the available methodology in this area is changing. Chapter 8 develops the area of longitudinal data analysis with multivariate response variables. This is a relatively new area with much potential for helping to understand how variables interact with each other within subjects across time. The Appendix lists FORTRAN subroutines for readers who would like to develop their own programs. While these routines have all been extracted from working programs, there is no guarantee that the routines are error free. This is particularly true of the last multivariate subroutine which is still in the developmental stage. This routine is listed in the hope that it will save time for some programmers, and, perhaps, encourage some people to write multivariate programs by simply giving them a start.

I would like to thank Professor Ulf Grenander of Brown University who steered me towards a career in statistics during my Ph.D. studies, Professor Emeritus David B. Duncan from the Johns Hopkins University who, as a friend and colleague, taught me much of what I know about statistics, and my friend and colleague at the University of Colorado Medical School, Professor Gary O. Zerbe, who introduced me to longitudinal data analysis, and from whom I learned much about this subject. Several students read drafts of the manuscript and made many helpful suggestions as well as finding errors. In particular, I would like to thank Luanne L. Esler, Becki Bucker, Yiming Zhang and David A. Young. Finally, I would like to thank Karen Kafadar, Ph.D. Fellow, Biometry Branch, Division of Cancer Prevention and Control, National Cancer Institute, for reading much of the manuscript and for making many constructive suggestions. It is always nice to have an accomplished statistician with a good command of the English language help with a manuscript.

This monograph has been typeset using LaTeX, and the figures have been drawn using PiCTeX, and was partially supported by the National Institute of General Medical Studies, Grant Number GM38519.

Richard H. Jones
Denver, Colorado

CHAPTER 1

Introduction

1.1 A simple example

In a typical medical application, subjects are assigned randomly to two or more treatment groups, one of which may be a placebo or control group. As an example, the treatment may be a medication for hypertension. The subjects are then followed over time, their blood pressure measured during visits to a clinic. In a balanced design, every subject has his or her blood pressure measured at equally spaced intervals of, say, one week until a given number of observations is obtained. The statistical question of interest is whether the mean blood pressures for the two groups are significantly different. Assuming that the responses to the treatment or placebo are statistically independent from subject to subject, a simple solution to this problem exists in the balanced case. The mean blood pressure for each subject can be calculated and the two groups compared using the nonparametric Wilcoxon rank sum test or the equivalent Mann Whitney test. If these means appear normally distributed from subject to subject, the two sample t-test can be used. The mathematical model for this simple application, for subjects in group 1, is

$$y_{i(1)j} = \beta_1 + \gamma_{i(1)} + \epsilon_{i(1)j},$$

and for subjects in group 2,

$$y_{i(2)j} = \beta_1 + \beta_2 + \gamma_{i(2)} + \epsilon_{i(2)j}. \tag{1.1}$$

The notation $i(k)$ denotes a subject in group k, and j denotes the observation on the subject. The notation $i(k)$ is standard notation when subjects are *nested* within groups. This model contains both fixed and random effects. The fixed effects are β_1 and β_2. β_1 is the population mean blood pressure of the subjects in group 1, and β_2 is the difference between the population means of group

2 and group 1. Assuming that there is a random distribution of mean blood pressure from subject to subject in the population, the mean blood pressure for a given subject is the value obtained by averaging many blood pressure measurements on this subject. The subject to subject variation in mean blood pressure is denoted by the random effect $\gamma_{i(k)}$. The variance of the random effect $\gamma_{i(k)}$ is often referred to as the between subject component of variation. The second random effect is $\epsilon_{i(k)j}$ and its variance is referred to as the within subject component of variation. The within subject random effect can consist of measurement error plus the actual variability of the subject's blood pressure.

Fig. 1.1 shows three sets of simulated data with both a random subject effect and random measurement error. The solid lines denote the population mean level and the dashed lines denote the individual subject's mean level. The observations vary randomly about the subject's mean level. Notice that in the second simulation all but one of the observations are below the group mean, and in the third simulation most of the observations are above the group mean. The random subject effect is causing correlation within each subject's observations, and the correlation between any two observations on a subject is the same. This constant correlation structure within the subjects is called compound symmetry, and is discussed in section 1.5.

The statistical hypothesis of interest in model (1.1) is

$$H_0 : \beta_2 = 0,$$

which states that the population means for the two groups are the same so that the two treatments do not differ in their effect on blood pressure. The results of a hypothesis test is either 'β_2 is significantly different from zero', or 'β_2 is not significantly different from zero.' Additional information can be conveyed by including a 'P-value'. A P-value is the probability that the estimate of β_2 would be farther from zero than the actual estimated value if the true value of β_2 were zero. The result of a significance test, together with a P-value, still do not give the reader a clear indication of both the magnitude of any difference that may exist between the results of the two treatments and the statistical variability of the estimate. When presenting the results of a statistical analysis, it is more informative to present a confidence interval for an estimated parameter than to present the result of a hypothesis test.

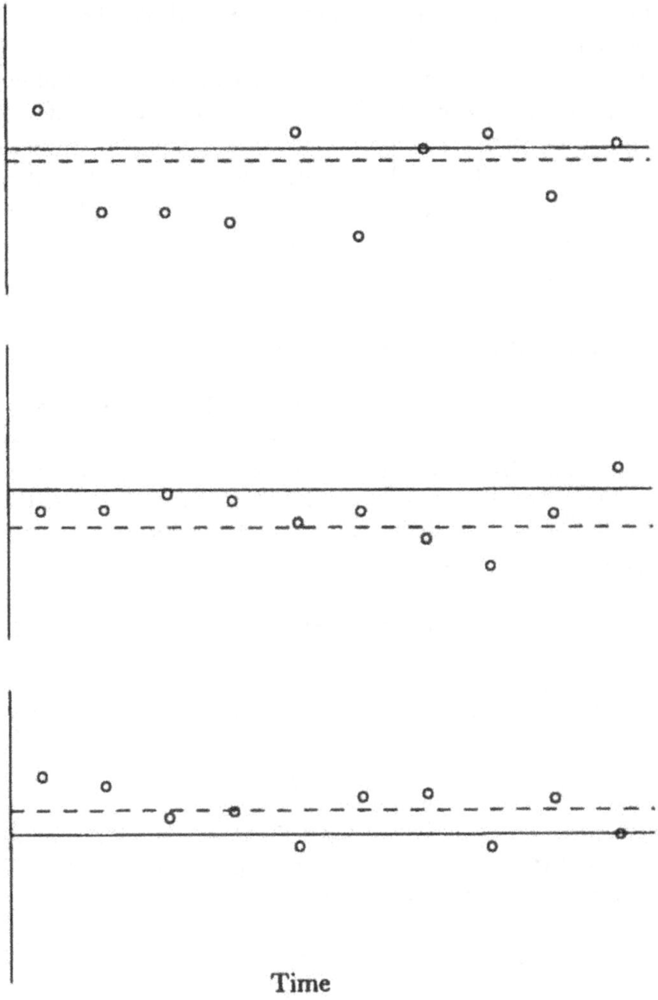

Fig. 1.1 *Three independent simulated sets of observations with random subject effect and random measurement error. The solid line is the group or population mean and the dashed line is the subjects's mean.*

Using $\hat{\beta}_2$ as the estimate of β_2, more information is conveyed by reporting $\hat{\beta}_2$ and the standard error of $\hat{\beta}_2$ than by reporting the results of a hypothesis test. An estimate with its standard error can be easily converted into a confidence interval.

The above example illustrates the nature of longitudinal data since each subject is observed more than once across time. Longitudinal data analysis studies a group or groups of subjects who are followed across time. Often a time response curve, sometimes called a *growth curve*, is of interest. In this example, the time response curve is a horizontal line that may be at a different level for each group.

1.2 Repeated measures experiments

Model (1.1) is very similar to a repeated measures experiment (Winer 1971); however, there are differences. In a classical single group repeated measures experiment, a group of subjects is given several treatments in an order that is different for different subjects, and a is response measured for each treatment. The mathematical model is,

$$y_{ij} = \beta_j + \gamma_i + \epsilon_{ij}. \tag{1.2}$$

Here, β_j is the mean effect for treatment j, and γ_i is a random subject effect for subject i. If there are n treatments, the hypothesis

$$H_0: \beta_1 = \beta_2 = \cdots = \beta_n,$$

tests whether response to the treatments are all the same. Rejection of this hypothesis is said to be a significant treatment effect. The reduced model in this case is

$$y_{ij} = \beta + \gamma_i + \epsilon_{ij},$$

which contains only a single coefficient for the single fixed treatment effect.

Rather than performing the hypothesis test in this example, confidence intervals can be calculated for the difference between pairs of β's of interest. For example, a confidence interval for the difference $\beta_3 - \beta_1$ would give a visual indication of whether treatment 3 and treatment 1 are different as well as an indication of the variability of the estimates. Looking at differences between regression coefficients is a special case of looking at *contrasts* which are linear combinations of regression coefficients.

Although time enters into this experiment, it is treatments that are of interest. Since different treatments are given at different times, care must be taken that the treatment times are far enough apart so that there is no carryover effect from one treatment to the next. By varying the order of treatments from subject to subject, the effect of treatment ordering is minimized.

In a two group repeated measures experiment, in addition to the treatment effect, there is also a group effect which is a vertical shift between groups, and a group by treatment interaction which is a different shaped time response for the two groups (Jones, 1985a).

1.3 Longitudinal data

Classical longitudinal data analysis is based on balanced designs where every subject is measured at the same time points with no missing observations. If the number of subjects is large relative to the number of observations per subject, multivariate analysis methods which assume a general covariance structure for the observations taken on the same subject can be used. Rao (1959) introduced the idea of stochastic or random parameters. This approach was developed further by Potthoff and Roy (1964), Rao (1965) and Grizzle and Allen (1969). These methods are quite powerful when balanced data are available.

Another approach to longitudinal data analysis is to calculate summary measures for each subject and then analyze the summary measures across subjects (see, for example, Matthews, et al., 1990). Suppose, for example, that there are two groups of subjects, and each subject's observations can be fit approximately by a straight line. One group receives a placebo and the other group receives a treatment, and the treatment is expected to affect the slopes of the lines. A straight line can be fit to each subject's data by simple linear regression, and the slopes compared between the two groups by using a two sample t-test or the corresponding nonparametric Wilcoxon rank sum test. Since data from different subjects are assumed to be independent, the slopes will be independent, and this is a proper analysis. The problem of correlation between observations within subjects is avoided. Using summary measures does require that the data be balanced or nearly balanced so that the summary measures can be assumed to be random samples from some distribution. For example, unbalanced data will cause the variances of the summary measures to be different for different subjects.

There are many possible complications in the analysis of longitudinal data, one of which is missing observations. As a result, different subjects are measured a different number of times, the mean blood pressure for the different subjects will have different variances which violates the assumptions necessary for the two sample t-test to be valid. Many computer programs cannot handle missing observations and simply eliminate from the analysis a subject with one or more missing observations.

Time cannot be randomized. The order of applying treatments can be randomized. When time response is of primary interest, there is the possibility of *serial correlation*. In the blood pressure example, the usual assumption is that repeated observations on each subject are randomly distributed about that subject's mean level. If observations are taken at time intervals that are too close together, an observation that is above the subject's mean level is more likely when the previous observations is above the mean level. This is serial correlation and it tends to die away as the time interval increases. Serial correlation is part of the error structure, and, if it exists, it must be modelled for a proper analysis.

Another possible complication is unequally spaced observations. It is not always possible to have patients return to a clinic at equally spaced intervals. This may not be a problem in a simple situation where there is no serial correlation so that the observations are statistically independent and each subject is measured the same number of times. However, if a time response curve is of interest, or if some of the observations are close enough together so that there is significant serial correlation, unequally spaced observations can cause problems. The serial correlation must be modelled in order to obtain correct tests of hypotheses or correct confidence intervals.

This book concentrates on models where the errors have Gaussian distributions. There has been much work in recent years on longitudinal data with non-Gaussian distributions. For example, Stiratelli, Laird and Ware (1984) consider serial observations with binary response. Quasi-likelihood methods (Wedderburn, 1974; Nelder and Pregibon, 1987; McCullagh and Nelder, 1989) and

Table 1.1 *A two group experiment with longitudinal data*

	Time (min)						
	0	5	10	20	30	45	60
Group 1	100	90	100	70	36	50	28
	100	83	97	60	83	83	97
	70	83	77	21	36	53	53
	77	36	36	36	36	36	28
	100	61	99	83	97	83	100
Group 2	90	28	44	33	36	44	36
	83	14	28	28	21	28	36
	69	36	28	21	14	21	21
	100	77	44	28	44	14	
	61	53	53	44	44	28	44

generalized estimating equations (Zeger and Liang, 1986; Liang and Zeger, 1986) provide a way to relax the assumption of Gaussian errors to distributions from the exponential family.

1.4 An example

As a faculty member in a biostatistics section of a department of preventive medicine and biometrics in a school of medicine, I consult with many researchers who have problems involving longitudinal data. Some of these researchers do not realize that the problem is one of longitudinal data. The person may or may not have experience with analysis of repeated measures experiments. One day a medical student appeared in my office with a 'statistics problem'. He explained the medical problem and the design of the experiment in detail and finally produced the data that are shown in Table 1.1. Each row consists of observations taken over time on the same subject. The two groups have different treatments. The problem is whether the two groups are responding the same or differently over time. I mentioned that there appeared to be a missing observation. He said 'oh yes, we didn't get that one observation on that subject'. I asked if he would feel comfortable if that subject or that time were removed from the analysis. He said that a lot of time and effort went into the collection of each observation and that he would prefer to use all the data if possible. I pointed out that this could be set up in SAS/STAT (SAS, 1985) using PROC GLM, and told him that I had a program that could handle this situation and even include serial correlation in the model if necessary, and I would try it out for him over the week-end. The newer SAS PROC MIXED is now available for handling unbalanced repeated measures or longitudinal data analysis.

The first step when examining a new data set is to *plot the data*. Individual curves and group mean curves can be plotted. The individual curves for the two groups are shown in Fig. 1.2.

The group mean curves with plus and minus one standard error calculated from the four or five observations available for the group at the time point are shown in Fig. 1.3. The standard errors are highly variable since the estimates have only four degrees of freedom (three in the case of Group 2 at 60 minutes). Group 1 may have higher standard deviations than Group 2 and the curves may or may not have a different shape. It also appears as if Group 1 may have larger standard errors than Group 2.

There are several mathematical models that can be fit to these data. The 'saturated model' fits a different mean to every time point for every group,

$$y_{i(k)j} = \beta_{jk} + \gamma_{i(k)} + \epsilon_{i(k)j}, \tag{1.3}$$

where k takes the values 1 and 2 denoting the group to which the subject belongs. There are 14 β_{jk}'s, for $j = 1$ to 7. This saturated model contains the time effect, the group effect and the group by time interaction. A reduced model with the group by time interaction removed is, for group 1,

$$y_{i(1)j} = \beta_{j1} + \gamma_{i(1)} + \epsilon_{i(1)j},$$

and for group 2,

$$y_{i(2)j} = \beta_{j1} + \beta_2 + \gamma_{i(2)} + \epsilon_{i(2)j}.$$

In this model, the shape of the curve for group one is arbitrary since there are 7 β_{j1}'s, and the two curves have a vertical shift of β_2. This vertical shift is the group effect. To test for a time effect, the null hypothesis is

$$H_0: \beta_{11} = \beta_{21} = \cdots = \beta_{71}.$$

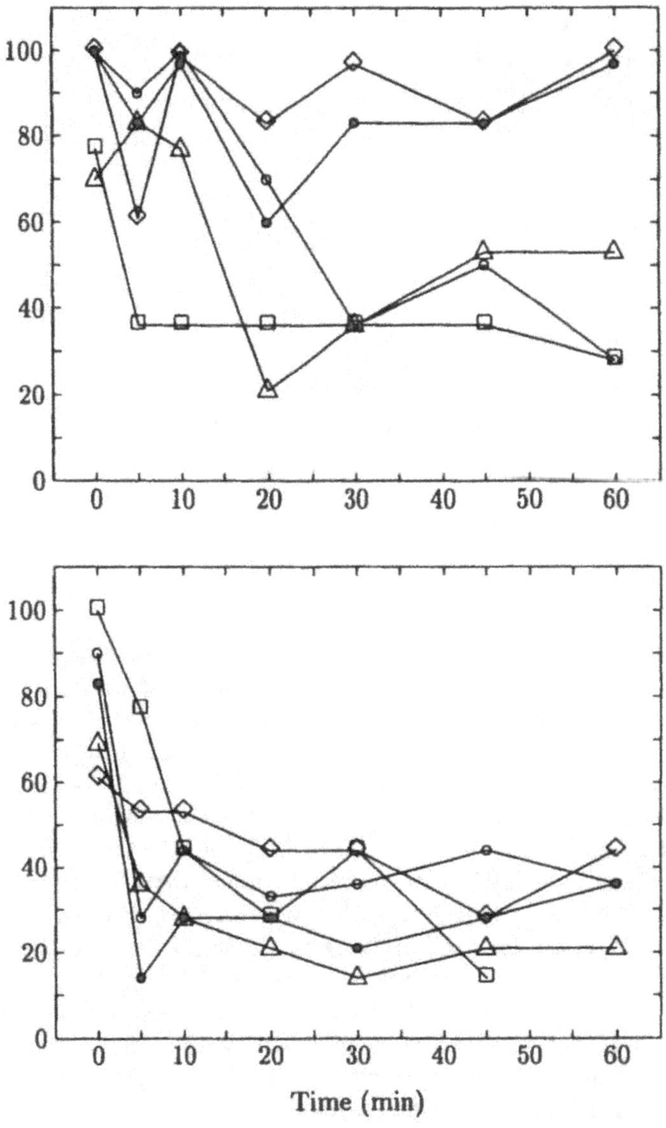

Fig. 1.2 *Individual curves. Group 1 above, Group 2 below.*

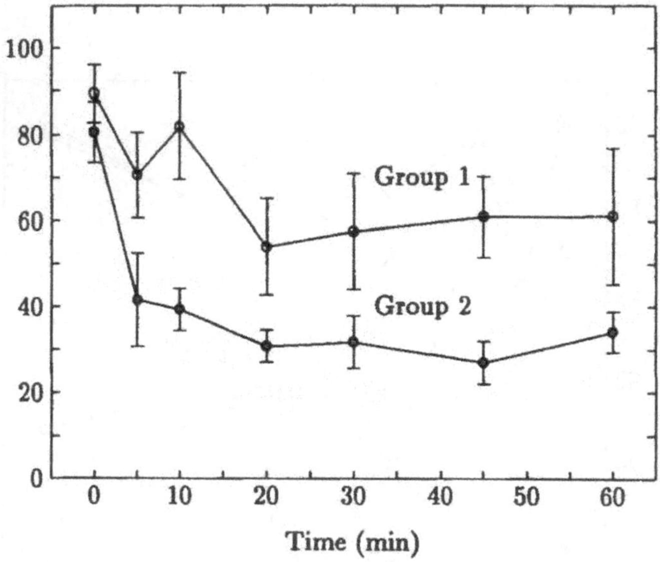

Fig. 1.3 *Group mean curves with plus and minus one standard error.*

To carry out hypothesis tests, there are two variances, σ_γ^2 and σ_ϵ^2, that need to be estimated. The details of the analysis are given in section 2.10, page 50. A summary of the results is that the data do not show significant serial correlation when random subject effects are included in the model. There is a significant time effect, indicating that the curves are not constant over time, and a significant group effect, indicating that the two group curves are shifted vertically with respect to each other. There is no significant group by time interaction, indicating that the shape of the two curves are the same. The result of fitting the reduced model where the two group curves have the same arbitrary shape, with one curve shifted vertically with respect to the other, is shown in Fig. 1.4. The standard error bars are now the same height except for Group 2 at 60 minutes which is slightly larger, although this difference cannot be seen by eye.

The presentation of this example at this stage is designed to demonstrate some of the problems that can be encountered in longitudinal data analysis. The details of the estimation procedure will be developed later. The important points at this stage are:

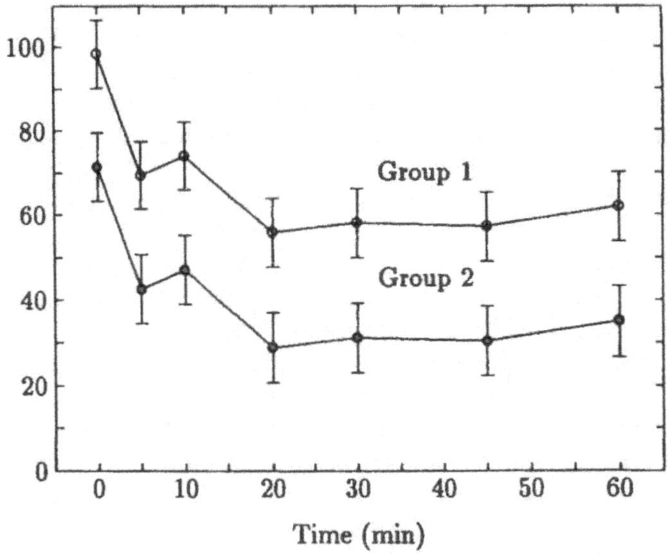

Fig. 1.4 *Results of fitting the mathematical model to the data.*

1. Plot the data as in Fig. 1.2. Plotting individual profile curves gives a visual picture of possible models for the data and the variability.

2. If the data are collected at the same time points for each subject, plot the group mean curves with error bars as in Fig. 1.3. Care must be taken not to draw inferences from these plots, but to use them only as an aid for model selection and an indication of possible variance heterogeneity.

3. Consider mathematical models that include random subject effects and possibly serial correlation. It may be that inclusion of one or the other, or perhaps both, of these effects gives an improved fit to the data (Jones, 1990).

1.5 Balanced compound symmetry

1.5.1 Maximum likelihood estimation

This section presents the details of maximum likelihood estimation in the balanced compound symmetry case. The mathematical manipulations of matrices that are necessary to obtain likelihoods for longitudinal data problems are developed. The complications that arise from unbalanced data are discussed. There are two basic approaches to calculating likelihoods developed in this book. This first approach could be called the direct or 'brute force' approach. Later, the Kalman filter will be introduced as an alternative method for calculating likelihoods.

In all but the most simple situations, nonlinear optimization routines must be used to obtain maximum likelihood estimates. When using a nonlinear optimization routine, a subroutine must be written that calculates the likelihood function for given values of the unknown parameters. The nonlinear optimization program then searches for the values of the parameters that maximize the likelihood function or, equivalently, minimize -2 ln likelihood, where ln denotes the natural logarithm to the base e. If the search is successful, maximum likelihood estimates are obtained. As a word of warning before diving into likelihood calculations, there can be problems. If the model is over parameterized relative to the number of observations available, the -2 ln likelihood may be very flat in places, and may not have a unique minimum.

Model (1.1) can be written as a single equation by defining an indicator variable for each subject that indicates the subject's group. Let

$$x_{i(k)j} = \begin{cases} 0 & \text{if } k = 1 \\ 1 & \text{if } k = 2 \end{cases},$$

The $i(k)j$ subscript indicates that subjects i are nested within groups k, and j indicated the measurements on subject i. Equation (1.1) can now be written

$$y_{i(k)j} = \beta_1 + \beta_2 x_{i(k)j} + \gamma_{i(k)} + \epsilon_{i(k)j}, \qquad (1.4)$$

where $i(k) = 1, \cdots, m_k$, and m_k is the number of subjects in group k, and $j = 1, \cdots, n_i$, where n_i is the number of observations on subject i.

Under the following assumptions, maximum likelihood estimates can be obtained in closed form. The assumptions are:

1. The $\gamma_{i(k)}$ are independent and normally (Gaussian) distributed random variables with mean zero and variance σ_γ^2 written

$$\gamma_{i(k)} \sim iN(0, \sigma_\gamma^2),$$

i.e., the mean blood pressures for different subjects are independent from subject to subject, and the distribution across subjects is Gaussian.

2. The $\epsilon_{i(k)j}$ are independent and normally distributed random variables that are independent of the $\gamma_{i(k)}$ and have mean zero and variance σ_ϵ^2,

$$\epsilon_{i(k)j} \sim iN(0, \sigma_\epsilon^2),$$

i.e., the repeated measurements on each subject are independent and normally distributed about the subject's own mean, $\beta_1 + \beta_2 x_{i(k)j} + \gamma_{i(k)}$.

Based on these assumptions, the variance of a single observation is

$$\text{var}\left(y_{i(k)j}\right) = \text{var}\left(\gamma_{i(k)} + \epsilon_{i(k)j}\right) = \sigma_\epsilon^2 + \sigma_\gamma^2,$$

since β_1 and β_2 are fixed (non-random) parameters, and $x_{i(k)j}$ is a non-random indicator variable. The covariance between any two different observations taken on the same subject (observation j and observation j') is

$$\text{cov}\left\{\left(\gamma_{i(k)} + \epsilon_{i(k)j}\right), \left(\gamma_{i(k)} + \epsilon_{i(k)j'}\right)\right\} = \sigma_\gamma^2,$$

since $\epsilon_{i(k)j}$ and $\epsilon_{i(k)j'}$ are uncorrelated. This assumes that σ_γ^2 is the same for all groups. In some applications, σ_γ^2 may be different for different groups. The covariance matrix for the observations on each subject is

$$\mathbf{V} = \begin{bmatrix} \sigma_\epsilon^2 + \sigma_\gamma^2 & \sigma_\gamma^2 & \sigma_\gamma^2 & \cdots & \sigma_\gamma^2 \\ \sigma_\gamma^2 & \sigma_\epsilon^2 + \sigma_\gamma^2 & \sigma_\gamma^2 & \cdots & \sigma_\gamma^2 \\ \sigma_\gamma^2 & \sigma_\gamma^2 & \sigma_\epsilon^2 + \sigma_\gamma^2 & \cdots & \sigma_\gamma^2 \\ \vdots & \vdots & \vdots & & \\ \sigma_\gamma^2 & \sigma_\gamma^2 & \sigma_\gamma^2 & \cdots & \sigma_\epsilon^2 + \sigma_\gamma^2 \end{bmatrix}. \quad (1.5)$$

Dividing each element by the product of the square roots of the corresponding diagonal elements, in this case $\sigma_\epsilon^2 + \sigma_\gamma^2$, produces the corresponding correlation matrix

$$\begin{bmatrix} 1 & \rho & \rho & \cdots & \rho \\ \rho & 1 & \rho & \cdots & \rho \\ \rho & \rho & 1 & \cdots & \rho \\ \vdots & \vdots & \vdots & & \\ \rho & \rho & \rho & \cdots & 1 \end{bmatrix}.$$

Covariance and correlation matrices of this form are said to have compound symmetry, and

$$\rho = \frac{\sigma_\gamma^2}{\sigma_\epsilon^2 + \sigma_\gamma^2} \tag{1.6}$$

is called the intraclass correlation coefficient. The intraclass correlation coefficient varies over the range zero to one. The case $\rho = 0$ occurs when $\sigma_\gamma^2 = 0$ so that $\gamma_{i(k)}$ can be deleted from the model. There is no between subject component of variation, and all observations would be independent observations of the group mean. This reduces the problem to simple regression. The case $\rho = 1$ occurs when $\sigma_\epsilon^2 = 0$ so that every measurement on a subject would be exactly the same and nothing is gained by taking multiple observations on a subject. The problem reduces to a two-sample t-test with a single observation on each subject. This case does not occur in practice since it implies that there is no measurement error and no intrasubject variability; however, it is possible for the intraclass correlation to be close to 1 indicating that the within subject variation is very small relative to the between subject variation.

The interpretation of the intraclass correlation sometimes is confusing to statisticians who do not have experience with repeated measures experiments. The constant correlation between measurements taken on the same subject is produced by the random subject effect. The actual within subject errors are independent. The random subject effect allows a given subject's mean to be above or below the group mean by a random amount. The subject has a higher or lower mean blood pressure than the group mean which produces the intraclass correlation. In practice, if the correlation between observations within a subject tends to be higher when the observations are closer together, this situation would be modelled as intraclass correlation (a random subject effect in the model) plus serial correlation.

The likelihood function is obtained by first writing the joint probability distribution of the observations for one subject in matrix form. Since different subjects are independent, the joint probability distribution for all the subjects is the product of the individual probability distributions. The random variables in the joint probability distribution are replaced by the actual observations, and the resulting function, the likelihood function, is considered to be a function of the unknown parameters. The values of the unknown parameters that maximize the likelihood function are the maximum likelihood estimates of the parameters. An intuitive interpretation is that the maximum likelihood estimates of the parameters are the values of the parameters that are most likely to have given the data that were actually observed.

In order to put the model in matrix form, write the observations for subject i in group k as a column vector of length n_i, $\mathbf{y}_{i(k)}$. Assume, for the moment, that the fixed effects part of the model is a constant mean which is different for each group, μ_k. The joint probability distribution of the observations for one subject is

$$f(\mathbf{y}_{i(k)}) = \frac{1}{(2\pi)^{\frac{n}{2}}|\mathbf{V}|^{\frac{1}{2}}} e^{-\frac{1}{2}(\mathbf{y}_{i(k)}-\mu_k\mathbf{1})'\mathbf{V}^{-1}(\mathbf{y}_{i(k)}-\mu_k\mathbf{1})},$$

where $\mathbf{1}$ is a column vector of ones, $|\mathbf{V}|$ denotes the determinant of the matrix $\mathbf{V}$ in equation (1.5), and $\mathbf{V}^{-1}$ denotes its inverse. The 'prime' denotes the transposed vector so that $(\mathbf{y}_{i(k)}-\mu_k\mathbf{1})'$ is a row vector and the exponent is a scalar, a quadratic form consisting of a row vector premultiplying a symmetric matrix with the same vector as a column vector postmultiplying the matrix.

The likelihood is the product of these functions over all the subjects with $\mathbf{y}_{i(k)}$ replaced by the data and considered to be a function of the four unknown parameters, μ_1, μ_2, σ^2_ϵ and σ^2_γ,

$$L(\mu_k, \sigma^2_\epsilon, \sigma^2_\gamma) = \prod_{i(k)} \frac{1}{(2\pi)^{\frac{n}{2}}|\mathbf{V}|^{\frac{1}{2}}} e^{-\frac{1}{2}(\mathbf{y}_{i(k)}-\mu_k\mathbf{1})'\mathbf{V}^{-1}(\mathbf{y}_{i(k)}-\mu_k\mathbf{1})}.$$

In this equation, $\prod_{i(k)}$ denotes the product over all the subjects. Maximum likelihood estimates of the parameters are obtained by finding the values of the parameters that maximize this function subject to the constraints that the parameters are within the parameter space,

$$\sigma^2_\gamma \geq 0$$

$$\sigma^2_\epsilon \geq 0.$$

It is often easier to find the maximum of the log of a function that consists of products than to find the maximum of the function itself. Both maxima are at the same values of the parameters. When numerical methods are used, most programs find a minimum rather than a maximum. Asymptotic likelihood ratio tests for testing hypotheses when fitting various models to a set of data are based on changes in −2 ln likelihood. For these reasons, it is convenient to work with −2 ln likelihood which will be denoted by ℓ,

$$\ell(\mu_k, \sigma_\epsilon^2, \sigma_\gamma^2) = \sum_{i(k)} [n \ln(2\pi) + \ln|\mathbf{V}| + (\mathbf{y}_{i(k)} - \mu_k \mathbf{1})' \mathbf{V}^{-1} (\mathbf{y}_{i(k)} - \mu_k \mathbf{1})] \tag{1.7}$$

where the summation is over all subjects in all groups.

The minimization of −2 ln likelihood equations is often simplified by concentrating out some of the parameters. ℓ is differentiated with respect to one of the parameters and the result set equal to zero. The resulting equation is solved for one parameter in terms of the other parameters, and the result substituted back into ℓ to give a new expression for ℓ with one less parameter. This is referred to as −2 ln likelihood concentrated with respect to the parameter that has been substituted out of the equation. In some simple cases, it is possible to differentiate ℓ with respect to all the parameters in the model and calculate the maximum likelihood estimates in closed form. In many practical situations, only a few parameters can be concentrated out of −2 ln likelihood and one must solve for the remaining parameters using nonlinear numerical optimization programs (Dennis and Schnabel, 1983).

Taking the partial derivative of equation (1.7) with respect to μ_1 and setting the result equal to zero,

$$\frac{\partial \ell}{\partial \mu_1} = \sum_{i(1)} \frac{\partial}{\partial \mu_1} (\mathbf{y}_{i(1)} - \mu_1 \mathbf{1})' \mathbf{V}^{-1} (\mathbf{y}_{i(1)} - \mu_1 \mathbf{1}) = 0, \tag{1.8}$$

gives the equation to be solved for μ_1. A similar equation is obtained for μ_2. Notice that the sum is over only the subjects in the corresponding group since the terms for the other group do not contain the parameter of interest. Expanding out (1.8) gives

$$\sum_{i(1)} \frac{\partial}{\partial \mu_1} (\mathbf{y}_{i(1)}' \mathbf{V}^{-1} \mathbf{y}_{i(1)} - \mu_1 \mathbf{y}_{i(1)}' \mathbf{V}^{-1} \mathbf{1} - \mu_1 \mathbf{1}' \mathbf{V}^{-1} \mathbf{y}_{i(1)} + \mu_1^2 \mathbf{1}' \mathbf{V}^{-1} \mathbf{1}) = 0.$$

Differentiating with respect to μ_1 gives

$$\sum_{i(1)} (-\mathbf{y}_{i(1)}' \mathbf{V}^{-1} \mathbf{1} - \mathbf{1}' \mathbf{V}^{-1} \mathbf{y}_{i(1)} + 2\mu_1 \mathbf{1}' \mathbf{V}^{-1} \mathbf{1}) = 0.$$

Since $\mathbf{y}'_{i(1)}\mathbf{V}^{-1}\mathbf{1}$ is a scalar, it is equal to its transpose $\mathbf{1}'\mathbf{V}^{-1}\mathbf{y}_{i(1)}$. If there are m_1 subjects in group 1, solving for μ_1 gives

$$\hat{\mu}_1 = \frac{\mathbf{1}'\mathbf{V}^{-1}\sum_{i(1)}\mathbf{y}_{i(1)}}{m_1\mathbf{1}'\mathbf{V}^{-1}\mathbf{1}}. \qquad (1.9)$$

The 'hat' on the μ indicates that this is an estimate which is a function of the observations rather than the true (unknown) value of the parameter. This is the well known result that is obtained using weighted least squares (Draper and Smith, 1981) and would be the maximum likelihood estimate of μ_1 if $\mathbf{V}$ were known. However, in the example being considered here, $\mathbf{V}$ depends on the two parameters σ_ϵ^2 and σ_γ^2.

The particular form of the covariance matrix in the compound symmetry example (1.5) can be written

$$\mathbf{V} = \sigma_\epsilon^2\mathbf{I} + \sigma_\gamma^2\mathbf{1}\mathbf{1}'. \qquad (1.10)$$

This matrix has the inverse

$$\mathbf{V}^{-1} = \frac{1}{\sigma_\epsilon^2}\left[\mathbf{I} - \frac{\sigma_\gamma^2\mathbf{1}\mathbf{1}'}{n\sigma_\gamma^2 + \sigma_\epsilon^2}\right], \qquad (1.11)$$

which can be verified by multiplying $\mathbf{V}$ by its inverse to demonstrate that the result is the identity matrix. Substituting this inverse into equation (1.9), and using the identities $\mathbf{1}'\mathbf{I} = \mathbf{1}'$, $\mathbf{1}'\mathbf{1} = n$, $\mathbf{I}\mathbf{1} = \mathbf{1}$ and $\mathbf{1}'\mathbf{y}_{i(1)} = \sum_{j=1}^n y_{i(1)j}$, the sum of the elements of the vector $\mathbf{y}_{i(1)}$, equation (1.9) simplifies to

$$\hat{\mu}_1 = \frac{1}{m_1 n}\sum_{i=1}^{m_1}\sum_{j=1}^{n} y_{i(1)j} = \bar{y}_{\cdot(1)\cdot}. \qquad (1.12)$$

This equation is the mean of all the observations in the group and is a natural estimate to use in any situation; however, it is the maximum likelihood estimation only in special cases.

To obtain -2 ln likelihood concentrated with respect to the μ_k, define the residual vectors from the group means

$$\tilde{\mathbf{y}}_{i(k)} = \mathbf{y}_{i(k)} - \hat{\mu}_k\mathbf{1}.$$

This equation is the vector of observations for subject i in group k with the mean of all observations of all the subjects in that group subtracted. Substituting back into equation (1.7) gives −2 ln likelihood concentrated with respect to the μ_k,

$$\ell(\sigma_\epsilon^2, \sigma_\gamma^2) = \sum_{i(k)} [n \ln(2\pi) + \ln |\mathbf{V}| + \tilde{\mathbf{y}}'_{i(k)} \mathbf{V}^{-1} \tilde{\mathbf{y}}_{i(k)}]. \tag{1.13}$$

This equation involves the determinant of $\mathbf{V}$. The determinant for the special case of compound symmetry (1.10) is given on page 67 in Rao (1973),

$$|\mathbf{V}| = (n\sigma_\gamma^2 + \sigma_\epsilon^2)(\sigma_\epsilon^2)^{n-1}. \tag{1.14}$$

When (1.11) and (1.14) are substituted into (1.13), the resulting expression contains only two functions of the observations,

$$T_T = \sum_{i(k)} \tilde{\mathbf{y}}'_{i(k)} \tilde{\mathbf{y}}_{i(k)} = \sum_{i(k)j} \tilde{y}^2_{i(k)j}, \tag{1.15}$$

which is the total sum of squares about the group means, and

$$T_B = \frac{1}{n} \sum_{i(k)} (\tilde{\mathbf{y}}'_{i(k)} \mathbf{1})(\mathbf{1}' \tilde{\mathbf{y}}_{i(k)}) = \frac{1}{n} \sum_{i(k)} \left(\sum_{j=1}^{n} \tilde{y}_{i(k)j} \right)^2 .$$

These are two sufficient statistics for the estimation of the two parameters σ_ϵ^2 and σ_γ^2. The sufficient statistics are two numbers calculated from the observations, and the maximum likelihood estimates of the parameters will be functions of these two numbers. In other words, the estimates depend on the data only through the two sufficient statistics. It is not always the case that the number of sufficient statistics equals the number of parameters. In some cases, the minimal set of sufficient statistics is the set of all the observations. In this balanced case there are two sufficient statistics for two parameters, the ideal situation.

Letting the number of subjects in the two groups be m_1 and m_2, with the total number of subjects being $m = m_1 + m_2$, and dropping the constant term $mn \ln(2\pi)$, −2 ln likelihood is

$$\begin{aligned} \ell(\sigma_\epsilon^2, \sigma_\gamma^2) &= m(n-1) \ln \sigma_\epsilon^2 + m \ln(n\sigma_\gamma^2 + \sigma_\epsilon^2) \\ &+ \frac{1}{\sigma_\epsilon^2} T_T - \frac{\sigma_\gamma^2}{\sigma_\epsilon^2 (n\sigma_\gamma^2 + \sigma_\epsilon^2)} n T_B. \end{aligned} \tag{1.16}$$

Taking the partial derivative of (1.16) with respect to σ_γ^2 and equating the result to zero gives

$$n\sigma_\gamma^2 + \sigma_\epsilon^2 = \frac{1}{m} T_B. \tag{1.17}$$

Taking the partial derivative of (1.16) with respect to σ_ϵ^2, equating the result to zero, and substituting $n\sigma_\gamma^2 + \sigma_\epsilon^2$ from equation(1.17) gives the following equation,

$$\sigma_\epsilon^2 = \frac{1}{m(n-1)} T_W, \tag{1.18}$$

where $T_W = T_T - T_B$ is the within subject sum of squares. Solving for σ_γ^2 from equations (1.17) and (1.18) gives

$$\sigma_\gamma^2 = \frac{1}{mn}\left(T_B - \frac{1}{n-1}T_W\right). \tag{1.19}$$

It is possible that this estimate could be negative falling outside the parameter space so that it is not the maximum likelihood estimate. In this case it is necessary to find the value of σ_ϵ^2 that minimizes ℓ on the boundary of the parameter space where $\sigma_\gamma^2 = 0$. To do this (1.16) must be rewritten with the constraint $\sigma_\gamma^2 = 0$,

$$\ell(\sigma_\epsilon^2, \sigma_\gamma^2 = 0) = mn \ln \sigma_\epsilon^2 + \frac{1}{\sigma_\epsilon^2} T_T.$$

This gives the estimate

$$\sigma_\epsilon^2 = \frac{1}{mn} T_T.$$

The maximum likelihood estimates are therefore:

- If $(n-1)T_B \geq T_W$, the usual case,

$$\tilde{\sigma}_\epsilon^2 = \frac{1}{m(n-1)} T_W$$

$$\tilde{\sigma}_\gamma^2 = \frac{1}{nm}\left(T_B - \frac{1}{n-1}T_W\right).$$

- If $(n-1)T_B < T_W$,

$$\tilde{\sigma}_\epsilon^2 = \frac{1}{mn} T_T$$

$$\tilde{\sigma}_\gamma^2 = 0.$$

Table 1.2 *The analysis of variance table for a two group repeated measures experiment*

Source	*d.f.*	*Mean Square*	*Expected Mean Square*
Subject	$m-2$	$T_B/(m-2)$	$\sigma_\epsilon^2 + n\sigma_\gamma^2$
Error	$m(n-1)$	$T_W/(mn-m)$	σ_ϵ^2

These full maximum likelihood estimates were derived by Herbach (1959).

The traditional estimates of σ_ϵ^2 and σ_γ^2 are based on the analysis of variance table shown in Table 1.2. There are $m-2$ degrees of freedom for subjects since this is a two group experiment with two group means estimated. Equating the mean squares to their expectations gives the method of moments estimates

$$\tilde{\sigma}_\epsilon^2 = \frac{1}{m(n-1)} T_W$$

$$\tilde{\sigma}_\gamma^2 = \frac{1}{n}\left(\frac{T_B}{m-2} - \frac{T_W}{m(n-1)}\right). \tag{1.20}$$

These equations differ from the maximum likelihood estimates in that T_B is divided by $m-2$ rather than m allowing for the two degrees of freedom lost by estimating the two means. This is similar to estimating the mean and variance of independent observations from a normal distribution from a sample of size n. The usual unbiased estimate divides the sum of squares of residuals by $n-1$, while the maximum likelihood estimate divides by n. Equation 1.20 can also produce negative estimates of $\tilde{\sigma}_\gamma^2$.

Thompson (1962) introduced the idea of restricted maximum likelihood for the purpose of obtaining unbiased estimates of variances and avoiding negative estimates of variances. If equation (1.20) produces a negative estimate, the estimates become

$$\tilde{\sigma}_\gamma^2 = 0,$$

$$\tilde{\sigma}_\epsilon^2 = \frac{T_T}{nm-2},$$

which differ from the maximum likelihood estimates by correcting for the two degrees of freedom lost by estimating two means. This later led to more general methods of obtaining unbiased estimates of variance referred to as restricted maximum likelihood or residual maximum likelihood (REML) (Robinson, D. L., 1987; Robinson, G. K., 1991).

An excellent reference for balanced repeated measures and longitudinal data analysis is Crowder and Hand (1990).

1.6 Missing observations

The closed form expressions obtained in the previous section depend on a balanced design where all subjects are observed at the same times and there are no missing observations. The example in section 1.4 contained a single missing observation. This section looks at the problems caused by missing observations and more general types of unbalanced designs.

It is important to consider the reason why there are missing observations. If the reason that an observation is missing is related to the response variables that are missed, the analysis will be biased. Rubin (1976) discusses the concept of data that are *missing at random* in a general statistical setting. Little and Rubin (1987) distinguish between observations that are *missing completely at random*, where the probably of missing an observation is independent of both the observed responses and the missing responces, and missing at random, where the probably of missing an observation is independent of the missed responses. Laird (1988), in a longitudinal data analysis setting, discusses observations that are missing completely at random, and uses the term *ignorable* missing data for observations that are missing at random. In this case, a likelihood based analysis with missing observations is valid. We will assume that the mechanism for missing observations is missing at random or ignorable.

Diggle (1989) discusses the problem of testing whether dropouts in repeated measures studies occur at random. Dropouts are a common source of missing observations, and it is important for the analysis to determine if the dropouts occur at random. Diggle's paper is discussed by Ridout (1991) with a response by Diggle.

In the previous section, if different subjects have different numbers of observations, the length of the vectors $\mathbf{y}_{i(k)}$ are different. If subject $i(k)$ has $n_{i(k)}$ observations, the covariance matrix for that subject will be an $n_{i(k)} \times n_{i(k)}$ matrix. Since these matrices are of different sizes, subscripts are necessary. These covariance matrices will now be denoted $\mathbf{V}_{i(k)}$. They are all of the form shown in equation (1.5); only the sizes are different. The vector of ones will be of different lengths so subscripts are again necessary, and the equation for −2 ln likelihood (1.7) becomes

$$\ell(\mu_k, \sigma_\epsilon^2, \sigma_\gamma^2) = \sum_{i(k)} [n_{i(k)} \ln(2\pi) + \ln |\mathbf{V}_{i(k)}| + (\mathbf{y}_{i(k)} - \mu_k \mathbf{1}_{i(k)})' \mathbf{V}_{i(k)}^{-1} (\mathbf{y}_{i(k)} - \mu_k \mathbf{1}_{i(k)})]. \quad (1.21)$$

Now it is not possible to obtain a closed form expression for $\hat{\mu}_1$ which does not contain σ_γ^2 and σ_ϵ^2 as in equation (1.12), so a closed form solution for the maximum likelihood estimates of the parameters is not possible.

All is not lost, however. In the unbalanced case, the maximum likelihood estimates of the group means depend only on the ratio of the two variances. Let

$$c^2 = \sigma_\gamma^2 / \sigma_\epsilon^2,$$

and denote the sum of the observations on subject $i(k)$ as

$$y_{i(k)\cdot} = \sum_{j=1}^{n_{i(k)}} y_{i(k)j}.$$

By differentiating (1.21) with respect to μ_k, and setting the result equal to zero, it can be shown that

$$\hat{\mu}_k = \frac{\sum_{i=1}^{m_k} \{y_{i(k)\cdot} / (1 + n_{i(k)} c^2)\}}{\sum_{i=1}^{m_k} \{n_{i(k)} / (1 + n_{i(k)} c^2)\}}. \quad (1.22)$$

As in the last section, define the residual vectors from the group means

$$\tilde{\mathbf{y}}_{i(k)} = \mathbf{y}_{i(k)} - \hat{\mu}_k \mathbf{1}_{i(k)} \quad (1.23)$$

Substituting (1.23) into (1.21) gives −2 ln likelihood concentrated with respect to the μ_k, as in equation (1.13),

$$\ell(\sigma_\epsilon^2, \sigma_\gamma^2) = n \ln(2\pi) + \sum_{i(k)} \left\{ \ln |\mathbf{V}_{i(k)}| + \tilde{\mathbf{y}}_{i(k)}' \mathbf{V}_{i(k)}^{-1} \tilde{\mathbf{y}}_{i(k)} \right\},$$

where n is the total number of observations on all subjects,

$$n = \sum_{i(k)} n_{i(k)}$$

Using equations (1.11) and (1.14),

$$\ell(\sigma_\epsilon^2, c^2) = n \ln(2\pi\sigma_\epsilon^2) + \sum_{i(k)} [\ln(1 + n_{i(k)}c^2)$$
$$+ \frac{1}{\sigma_\epsilon^2} \left\{ \tilde{\mathbf{y}}'_{i(k)}\tilde{\mathbf{y}}_{i(k)} - (c\tilde{y}_{i(k)\cdot})^2/(1 + n_{i(k)}c^2) \right\}]. \quad (1.24)$$

Equation (1.24) is the form of the likelihood reparameterized in terms of σ_ϵ^2 and c^2. Differentiating with respect to σ_ϵ^2 and equating to zero gives

$$\hat{\sigma}_\epsilon^2 = \frac{1}{n} \sum_{i(k)} \left\{ \tilde{\mathbf{y}}'_{i(k)}\tilde{\mathbf{y}}_{i(k)} - (c\tilde{y}_{i(k)\cdot})^2/(1 + n_{i(k)}c^2) \right\}. \quad (1.25)$$

Substituting (1.25) back into (1.24) gives −2 ln likelihood as a function of a single unknown parameter, c^2,

$$\ell(c^2) = n \ln(2\pi\hat{\sigma}_\epsilon^2) + \sum_{i(k)} \{\ln(1 + n_{i(k)}c^2)\} + n. \quad (1.26)$$

For a given value of c^2, the group means are estimated using equation (1.22), the residuals calculated from (1.23), the estimate of σ_ϵ^2 from (1.25), and −2 ln likelihood from (1.26).

Equations (1.22) to (1.26) demonstrate the general idea of using nonlinear optimization for maximum likelihood estimates. The linear fixed effect parameters, as well as one of the variance parameters, can be concentrated out of the likelihood. By searching over the parameter space for the remaining nonlinear parameters, the maximum likelihood estimates often can be found. The parameter space consists of all legal values of the parameters; e.g., negative variances are outside of the parameter space.

There are sometimes numerical problems when attempting to find the maximum likelihood estimates. The likelihood function may not have a unique maximum within the parameter space, so an optimization routine may converge to a local maximum rather than to the global maximum. To avoid converging to a local maximum, it helps to have some idea of the approximate location of the unknown parameters. Without this knowledge, programs should be run with various starting values for the unknown parameters. Consistent convergence to the same point gives confidence that the point of convergence is the global estimate. If some starting values converge to one point and other starting values converge to other points, the points with the higher likelihood (lower values of −2 ln likelihood) are the points of interest. When a model is over-parameterized, numerical problems are more likely to be encountered; therefore, starting with a simple model and working towards more complicated models is the usual method of choice. From the simpler models, initial guesses for the parameters of more complicated models often can be obtained.

In the above problem there is only one nonlinear parameter, c^2, which must be non-negative. Since there is only one non-linear parameter, it is possible to search for this parameter by hand by writing a program that calculates $\ell(c^2)$ for a given value of c^2. The user enters a value of c^2 and the program calculates and writes to the screen the value of $\ell(c^2)$. The program then returns to the beginning and asks for a new value of c^2 to be entered. If no prior information is available, the value of $c^2 = 0$ can be tried first, then a small positive value of c^2. It is possible that the maximum likelihood estimate is $c^2 = 0$. In this case, $\ell(c^2)$ will increase as c^2 moves away from zero. If $\ell(c^2)$ decreases as c^2 moves away from zero, larger values must be tried until $\ell(c^2)$ starts to increase. There are very efficient algorithms for a one dimensional search for a minimum (e.g. Press et al., 1986), but, even without such algorithms, a person can usually find a good approximation to a minimum in a few minutes time.

1.7 Exercises

1. Verify by multiplication that equation (1.11) is the inverse of $\mathbf{V}$.

2. Verify (1.17).

3. Verify (1.18).

4. Verify (1.22).

5. Write a program to evaluate $\ell(c^2)$ in equation (1.26), and use this program to find the approximate maximum likelihood

estimate of μ_1, μ_2, σ_ϵ^2 and σ_γ^2 for the data in Table 1.1 assuming that model (1.4) is correct for these data.

CHAPTER 2

A General Linear Mixed Model

2.1 The Laird-Ware Model

A very general linear mixed model for longitudinal data has been proposed by Laird and Ware (1982) based on the work of Harville (1974, 1976, 1977),

$$\mathbf{y}_i = \mathbf{X}_i\boldsymbol{\beta} + \mathbf{Z}_i\boldsymbol{\gamma}_i + \boldsymbol{\epsilon}_i, \tag{2.1}$$

where $\mathbf{y}_i$ is an $n_i \times 1$ column vector of the response variable for subject i, $\mathbf{X}_i$ is an $n_i \times b$ design matrix, $\boldsymbol{\beta}$ is a $b \times 1$ vector of regression coefficients assumed to be fixed, $\mathbf{Z}_i$ is an $n_i \times g$ design matrix for the random effects, $\boldsymbol{\gamma}_i$, which are assumed to be independently distributed across subjects with distribution, $\boldsymbol{\gamma}_i \sim N(0, \sigma^2\mathbf{B})$, where $\mathbf{B}$, for between, is an arbitrary covariance matrix. The within subject errors, $\boldsymbol{\epsilon}_i$, are assumed to be distributed $\boldsymbol{\epsilon}_i \sim N(0, \sigma^2\mathbf{W}_i)$, where $\mathbf{W}_i$, for within, is a covariance matrix which may be parameterized using a few parameters with a scale factor, σ^2, factored out. Often it is assumed that $\mathbf{W}_i = \mathbf{I}$, the identity matrix. Various possibilities for parameterizing the $\mathbf{W}_i$ matrices will be discussed later. For example, $\mathbf{W}_i$ may have the structure of a first order autoregression. The $\boldsymbol{\epsilon}_i$ are also independently distributed from subject to subject and independent of the $\boldsymbol{\gamma}_i$. Model (2.1) is very general since different subjects can have different numbers of observations as well as different observation times. The subscript i in the vector $\mathbf{y}_i$ and the matrices $\mathbf{X}_i$, $\mathbf{Z}_i$ and $\mathbf{W}_i$ indicates that these vectors and matrices are subject specific. The matrices $\mathbf{X}_i$ and $\mathbf{Z}_i$ are not necessarily of full rank.

Many common statistical models are special cases of equation (2.1). The repeated measures experiment (1.2) is one of the simplest examples. Consider the balanced case where each subject is given four treatments. The model for subject i is

$$\begin{bmatrix} y_{i1} \\ y_{i2} \\ y_{i3} \\ y_{i4} \end{bmatrix} = \begin{bmatrix} 1 & 0 & 0 & 0 \\ 0 & 1 & 0 & 0 \\ 0 & 0 & 1 & 0 \\ 0 & 0 & 0 & 1 \end{bmatrix} \begin{bmatrix} \beta_1 \\ \beta_2 \\ \beta_3 \\ \beta_4 \end{bmatrix} + \begin{bmatrix} 1 \\ 1 \\ 1 \\ 1 \end{bmatrix} \gamma_i + \begin{bmatrix} \epsilon_{i1} \\ \epsilon_{i2} \\ \epsilon_{i3} \\ \epsilon_{i4} \end{bmatrix},$$

so that $\mathbf{X}_i = \mathbf{I}$, the identity matrix, and $\mathbf{Z}_i = \mathbf{1}$, a column of ones, and the β's are the treatment means.

For the two group case (1.4), the model for subjects in group 1, assuming four observations per subject, is

$$\begin{bmatrix} y_{i(1)1} \\ y_{i(1)2} \\ y_{i(1)3} \\ y_{i(1)4} \end{bmatrix} = \begin{bmatrix} 1 & 0 \\ 1 & 0 \\ 1 & 0 \\ 1 & 0 \end{bmatrix} \begin{bmatrix} \beta_1 \\ \beta_2 \end{bmatrix} + \begin{bmatrix} 1 \\ 1 \\ 1 \\ 1 \end{bmatrix} \gamma_{i(1)} + \begin{bmatrix} \epsilon_{i(1)1} \\ \epsilon_{i(1)2} \\ \epsilon_{i(1)3} \\ \epsilon_{i(1)4} \end{bmatrix}, \quad (2.2)$$

and for subjects in group 2,

$$\begin{bmatrix} y_{i(2)1} \\ y_{i(2)2} \\ y_{i(2)3} \\ y_{i(2)4} \end{bmatrix} = \begin{bmatrix} 1 & 1 \\ 1 & 1 \\ 1 & 1 \\ 1 & 1 \end{bmatrix} \begin{bmatrix} \beta_1 \\ \beta_2 \end{bmatrix} + \begin{bmatrix} 1 \\ 1 \\ 1 \\ 1 \end{bmatrix} \gamma_{i(2)} + \begin{bmatrix} \epsilon_{i(2)1} \\ \epsilon_{i(2)2} \\ \epsilon_{i(2)3} \\ \epsilon_{i(2)4} \end{bmatrix}.$$

The β_1 is the mean of group 1, and $\beta_1 + \beta_2$ is the mean of group 2.

Jennrich and Schluchter (1986) fit a number of different models and different error structures to a data set used by Potthoff and Roy (1964). The data consist of measurements taken on 11 girls and 16 boys at ages 8, 10, 12 and 14 years of age. It is a balanced design with no missing observations. There are three different models for the means or fixed effects. Denoting the two groups as 1 and 2, the first model, for group 1 is

$$\mathbf{X}_i\boldsymbol{\beta} = \begin{bmatrix} 1 & 0 & 0 & 0 & 0 & 0 & 0 & 0 \\ 0 & 1 & 0 & 0 & 0 & 0 & 0 & 0 \\ 0 & 0 & 1 & 0 & 0 & 0 & 0 & 0 \\ 0 & 0 & 0 & 1 & 0 & 0 & 0 & 0 \end{bmatrix} \begin{bmatrix} \beta_1 \\ \beta_2 \\ \beta_3 \\ \beta_4 \\ \beta_5 \\ \beta_6 \\ \beta_7 \\ \beta_8 \end{bmatrix},$$

and, for group 2,

$$\mathbf{X}_i\boldsymbol{\beta} = \begin{bmatrix} 0 & 0 & 0 & 0 & 1 & 0 & 0 & 0 \\ 0 & 0 & 0 & 0 & 0 & 1 & 0 & 0 \\ 0 & 0 & 0 & 0 & 0 & 0 & 1 & 0 \\ 0 & 0 & 0 & 0 & 0 & 0 & 0 & 1 \end{bmatrix} \begin{bmatrix} \beta_1 \\ \beta_2 \\ \beta_3 \\ \beta_4 \\ \beta_5 \\ \beta_6 \\ \beta_7 \\ \beta_8 \end{bmatrix}.$$

These two equations represent what is often called the saturated model since each group can have a different and arbitrary growth curve. The number of coefficients is equal to the number of time points multiplied by the number of groups. The growth curve for group 1 is given by the coefficients β_1 to β_4, and the growth curve for group 2 is given by β_5 to β_8. There are any number of possible parameterizations for this model, and often a different choice gives a different interpretation and may be more convenient for hypothesis testing for the purpose of model selection. For example, another possibility for group 1 is

$$\mathbf{X}_i\boldsymbol{\beta} = \begin{bmatrix} 1 & 0 & 0 & 0 & 0 & 0 & 0 & 0 \\ 0 & 1 & 0 & 0 & 0 & 0 & 0 & 0 \\ 0 & 0 & 1 & 0 & 0 & 0 & 0 & 0 \\ 0 & 0 & 0 & 1 & 0 & 0 & 0 & 0 \end{bmatrix} \begin{bmatrix} \beta_1 \\ \beta_2 \\ \beta_3 \\ \beta_4 \\ \beta_5 \\ \beta_6 \\ \beta_7 \\ \beta_8 \end{bmatrix},$$

and for group 2,

$$\mathbf{X}_i\boldsymbol{\beta} = \begin{bmatrix} 1 & 0 & 0 & 0 & 1 & 0 & 0 & 0 \\ 0 & 1 & 0 & 0 & 0 & 1 & 0 & 0 \\ 0 & 0 & 1 & 0 & 0 & 0 & 1 & 0 \\ 0 & 0 & 0 & 1 & 0 & 0 & 0 & 1 \end{bmatrix} \begin{bmatrix} \beta_1 \\ \beta_2 \\ \beta_3 \\ \beta_4 \\ \beta_5 \\ \beta_6 \\ \beta_7 \\ \beta_8 \end{bmatrix}.$$

In this form the coefficients β_1 to β_4 still generate the growth curve for group 1, but now β_5 to β_8 are the differences between the two curves. The first time point for group two is $\beta_1 + \beta_5$. Testing the hypothesis

$$H_0: \beta_5 = \beta_6 = \beta_7 = \beta_8 = 0$$

tests whether the two curves are the same. If this hypothesis is true, the four coefficients, β_5 to β_8 reduce to a single coefficient. If we call this single coefficient β_5, this coefficient represents the group effect and there is no interaction. The coefficient β_5 in the reduced model is the vertical shift between the growth curve for the two groups.

The second model for the means is a straight line for the girls and a different straight line for the boys. This model has four fixed coefficients. One possibility for the fixed effects design matrix for this model is, for group 1,

$$\mathbf{X}_i\boldsymbol{\beta} = \begin{bmatrix} 1 & 8 & 0 & 0 \\ 1 & 10 & 0 & 0 \\ 1 & 12 & 0 & 0 \\ 1 & 14 & 0 & 0 \end{bmatrix} \begin{bmatrix} \beta_1 \\ \beta_2 \\ \beta_3 \\ \beta_4 \end{bmatrix}, \qquad (2.3)$$

and for group 2,

$$\mathbf{X}_i\boldsymbol{\beta} = \begin{bmatrix} 1 & 8 & 1 & 8 \\ 1 & 10 & 1 & 10 \\ 1 & 12 & 1 & 12 \\ 1 & 14 & 1 & 14 \end{bmatrix} \begin{bmatrix} \beta_1 \\ \beta_2 \\ \beta_3 \\ \beta_4 \end{bmatrix}. \qquad (2.4)$$

Here β_1 and β_2 are the intercept and slope for the line for group 1, and β_3 and β_4 are the differences between the intercepts and slopes for group 2 and group 1. The second column for the first group and the second and fourth column for the second group are the children's ages at the time of the observations.

The hypothesis that the two lines are parallel is

$$H_0: \beta_4 = 0,$$

which, if true, produces the third model for the means. For group 1,

$$\mathbf{X}_i\boldsymbol{\beta} = \begin{bmatrix} 1 & 8 & 0 \\ 1 & 10 & 0 \\ 1 & 12 & 0 \\ 1 & 14 & 0 \end{bmatrix} \begin{bmatrix} \beta_1 \\ \beta_2 \\ \beta_3 \end{bmatrix},$$

and for group 2,

$$\mathbf{X}_i\boldsymbol{\beta} = \begin{bmatrix} 1 & 8 & 1 \\ 1 & 10 & 1 \\ 1 & 12 & 1 \\ 1 & 14 & 1 \end{bmatrix} \begin{bmatrix} \beta_1 \\ \beta_2 \\ \beta_3 \end{bmatrix}.$$

The coefficient β_3 now is the vertical distance between the two parallel lines.

The above shows various possibilities with the fixed effects part of the model which relates to the population mean. There are also many possibilities for the random parts of the model relating to between subject random effects and within subject error structure. The first possibility, referred to by Jennrich and Schluchter (1986) as 'unstructured' (their models 1, 2 and 3), is that the covariance matrix for the observations on each subject is an arbitrary 4×4 covariance matrix. This error structure can be accomplished in equation (2.1) by setting one of the random terms to zero and letting the other term produce an arbitrary covariance matrix. For example, setting $\mathbf{W}_i = 0$ eliminates the $\boldsymbol{\epsilon}_i$ term, and if $\mathbf{Z}_i$ is set equal to a 4×4 identity matrix, the resulting error structure has the arbitrary covariance matrix $\sigma^2\mathbf{B}$. The other possibility is to set $\mathbf{B} = 0$ and let $\mathbf{W}_i$ be an arbitrary correlation matrix.

At the other extreme from an arbitrary error covariance matrix is an identity matrix for the error covariance matrix, which is Jennrich and Schluchter's model 8. In equation (2.1), this model can be obtained by setting $\mathbf{B} = 0$, which eliminates the term $\mathbf{Z}_i\boldsymbol{\gamma}_i$, and setting $\mathbf{W}_i = \mathbf{I}$. Their model 7 is compound symmetry which is left as an exercise.

Jennrich and Schluchter's model 6 is called 'random coefficients'. In this model, the fixed effects are a separate line for each group (model 2 above), but each subject has an intercept and slope that varies randomly about the group intercept and slope. The fixed effects part of the model is as in equations (2.3) and (2.4), and the between subject component of variance is

$$\mathbf{Z}_i\boldsymbol{\gamma}_i = \begin{bmatrix} 1 & 8 \\ 1 & 10 \\ 1 & 12 \\ 1 & 14 \end{bmatrix} \begin{bmatrix} \gamma_{i1} \\ \gamma_{i2} \end{bmatrix}.$$

The within subject errors are assumed to be uncorrelated with constant variance, $\boldsymbol{\epsilon}_i \sim N(0, \sigma^2\mathbf{I})$, i.e. $\mathbf{W}_i = \mathbf{I}$. γ_{i1} and γ_{i2} are the random deviations of the intercept and slope for subject i from the subject's group mean intercept and slope.

The introduction of random coefficient models begins to show the generality of the Laird-Ware model. Not only do the random effects allow a vertical shift in each subject's data, but they also allow a change in slope for each subject. These random deviations are usually assumed to have an arbitrary 2×2 covariance matrix across subjects, $\sigma^2\mathbf{B}$, since the random intercepts and slopes tend to be correlated. Also, this covariance matrix depends on the choice of the time origin. In some applications, random coefficient models help with modelling variance heterogeneity since the variance of the random coefficient model may increase or decrease with time (for an example see Jones, 1990).

Two other models considered by Jennrich and Schluchter have within subject error structures that are time series models. Model 5 is a first order autoregression where the between subject component is set equal to zero ($\mathbf{B} = 0$) and the within subject error correlation structure is of the form

$$\mathbf{W}_i = \begin{bmatrix} 1 & \rho & \rho^2 & \rho^3 \\ \rho & 1 & \rho & \rho^2 \\ \rho^2 & \rho & 1 & \rho \\ \rho^3 & \rho^2 & \rho & 1 \end{bmatrix}. \tag{2.5}$$

The autoregression coefficient ρ must be in the interval $-1 < \rho < 1$ for this correlation matrix to be positive definite.

Model 4 is called 'banded' and, again, $\mathbf{B} = 0$ and the within subject error correlation structure is of the form

$$\mathbf{W}_i = \begin{bmatrix} 1 & \rho_1 & \rho_2 & \rho_3 \\ \rho_1 & 1 & \rho_1 & \rho_2 \\ \rho_2 & \rho_1 & 1 & \rho_1 \\ \rho_3 & \rho_2 & \rho_1 & 1 \end{bmatrix}. \tag{2.6}$$

In this matrix, the three ρ's are different numbers. The model represents a serial correlation structure where the correlation depends only on the time difference between the observations. When correlations between observations on the same subject depend only on the time difference, the correlation structure is called *stationary*. The correlation matrix (2.5) is a special case of a stationary correlation structure. However, the ρ's may not be completely arbitrary since for certain values of the ρ's this matrix will not be positive definite. This banded structure with constants along every diagonal is also known as a Toeplitz matrix. Models 4 and 5 will be discussed in more detail later.

2.2 The likelihood function

The likelihood for the model (2.1) is similar to the likelihoods developed in Chapter 1; however, model (2.1) is much more general. The subscript k which is used in Chapter 1 to denote the group is no longer necessary. If there are several groups of subjects, this will be incorporated into the design matrix $\mathbf{X}_i$. The mean vector for subject i is now $\mathbf{X}_i\boldsymbol{\beta}$. Since the two random components have zero means, $\boldsymbol{\gamma}_i$ is uncorrelated with $\boldsymbol{\epsilon}_i$ and the two covariance matrices are $\text{cov}(\boldsymbol{\gamma}_i) = \sigma^2\mathbf{B}$ and $\text{cov}(\boldsymbol{\epsilon}_i) = \sigma^2\mathbf{W}_i$, the total covariance matrix for subject i is $\sigma^2(\mathbf{Z}_i\mathbf{B}\mathbf{Z}_i' + \mathbf{W}_i)$. Let

$$\mathbf{V}_i = \mathbf{Z}_i\mathbf{B}\mathbf{Z}_i' + \mathbf{W}_i.$$

Now the total covariance matrix for subject i is $\sigma^2\mathbf{V}_i$. Assuming that there are n_i observations for subject i, –2 ln likelihood is

$$\ell = \sum_i [n_i \ln(2\pi) + \ln|\sigma^2\mathbf{V}_i| + (\mathbf{y}_i - \mathbf{X}_i\boldsymbol{\beta})'(\sigma^2\mathbf{V}_i)^{-1}(\mathbf{y}_i - \mathbf{X}_i\boldsymbol{\beta})].$$

The σ^2 inside the determinant multiplies every element of the matrix $\mathbf{V}_i$, and since the determinant is the sum of products of n_i elements of the matrix,

$$\ln|\sigma^2\mathbf{V}_i| = \ln(\sigma^{2n_i}|\mathbf{V}_i|) = n_i \ln\sigma^2 + \ln|\mathbf{V}_i|.$$

Let $n = \sum_i n_i$ be the total number of observations on all the subjects. Then

$$\begin{aligned} \ell &= n\ln(2\pi) + n\ln\sigma^2 \\ &+ \sum_i \left[\ln|\mathbf{V}_i| + \frac{1}{\sigma^2}(\mathbf{y}_i - \mathbf{X}_i\boldsymbol{\beta})'\mathbf{V}_i^{-1}(\mathbf{y}_i - \mathbf{X}_i\boldsymbol{\beta})\right]. \end{aligned} \quad (2.7)$$

Differentiating ℓ with respect to σ^2 and setting the result equal to zero gives

$$\hat{\sigma}^2 = \frac{1}{n}\sum_i (\mathbf{y}_i - \mathbf{X}_i\boldsymbol{\beta})'\mathbf{V}_i^{-1}(\mathbf{y}_i - \mathbf{X}_i\boldsymbol{\beta}), \quad (2.8)$$

and substituting this back into (2.7) gives −2 ln likelihood concentrated with respect to σ^2,

$$\begin{aligned} \ell &= n\ln(2\pi) + n\ln\hat{\sigma}^2 + \sum_i(\ln|\mathbf{V}_i|) + n \\ &= n\ln(2\pi\hat{\sigma}^2) + \sum_i(\ln|\mathbf{V}_i|) + n. \end{aligned} \qquad (2.9)$$

By keeping the constant terms in the likelihoods, it is possible to compare results when using the complete and the concentrated forms of the likelihoods. It is also possible to compare results with published results such as Jennrich and Schluchter (1986) who do not drop constants from the likelihoods.

The unknown parameters in (2.9) are the β vector, the $\mathbf{B}$ matrix and any parameters that are used to parameterize the matrices $\mathbf{W}_i$. For given values of the $\mathbf{B}$ and $\mathbf{W}_i$ matrices, the estimate of β that minimizes ℓ is the solution of the *normal equations*,

$$\left(\sum_i \mathbf{X}_i'\mathbf{V}_i^{-1}\mathbf{X}_i\right)\hat{\beta} = \left(\sum_i \mathbf{X}_i'\mathbf{V}_i^{-1}\mathbf{y}_i\right), \qquad (2.10)$$

$$\hat{\beta} = (\sum_i \mathbf{X}_i'\mathbf{V}_i^{-1}\mathbf{X}_i)^{-1}(\sum_i \mathbf{X}_i'\mathbf{V}_i^{-1}\mathbf{y}_i). \qquad (2.11)$$

The word *normal* in normal equations has nothing to do with the Gaussian distribution, but is related to the word orthogonal. The theory of least squares can be developed using orthogonal projections (Watson, 1967). Equation (2.11) is the well known weighted least squares estimate when the error covariance matrix in linear regression is $\sigma^2\mathbf{V}_i$ (Draper and Smith, 1981, section 2.11). Substituting this estimate into (2.8) and the resulting $\hat{\sigma}^2$ into (2.9) gives −2 ln likelihood concentrated with respect to both σ^2 and β. The only nonlinear parameters remaining are the parameters in the covariance matrices.

2.2.1 Numerical evaluation

For $\mathbf{B}$ to be a proper covariance matrix, it is necessary for it to be non-negative definite. A matrix is non-negative definite if for any column vector of constants, $\mathbf{c}$,

$$\mathbf{c}'\mathbf{B}\mathbf{c} \geq 0. \qquad (2.12)$$

A matrix is positive definite if for any column vector of constants, $\mathbf{c}$, where all the elements are not 0, $\mathbf{c}'\mathbf{c} > 0$,

$$\mathbf{c}'\mathbf{B}\mathbf{c} > 0.$$

If a nonlinear optimization routine is allowed to search over arbitrary values of the elements of $\mathbf{B}$, some of the resulting matrices will not be non-negative definite so the parameters will not be in the parameter space. To prevent this problem, the matrix $\mathbf{B}$ is represented in a factored form,

$$\mathbf{B} = \mathbf{U}'\mathbf{U}, \tag{2.13}$$

where $\mathbf{U}$ is an upper triangular matrix, i.e. all the elements below the main diagonal are zero. If $\mathbf{B}$ is positive definite, (2.13) is the well known Cholesky factorization of the matrix $\mathbf{B}$. For any matrix $\mathbf{U}$, the resulting $\mathbf{B}$ will be non-negative definite, since, by substituting (2.13) into (2.12),

$$\mathbf{c}'\mathbf{B}\mathbf{c} = \mathbf{c}'\mathbf{U}'\mathbf{U}\mathbf{c} = (\mathbf{U}\mathbf{c})'(\mathbf{U}\mathbf{c}).$$

$\mathbf{U}\mathbf{c}$ is a column vector, and $(\mathbf{U}\mathbf{c})'(\mathbf{U}\mathbf{c})$ is the sum of squares of the elements of the vector; therefore,

$$(\mathbf{U}\mathbf{c})'(\mathbf{U}\mathbf{c}) \geq 0.$$

When using a nonlinear optimization program to find maximum likelihood estimates, it is necessary to write a subroutine to calculate −2 ln likelihood for given values of the nonlinear parameters. The optimization program then searches over values of the nonlinear parameters in an attempt to find the minimum of −2 ln likelihood. For subject i, the data consist of the vector $\mathbf{y}_i$ and the matrices $\mathbf{X}_i$ and $\mathbf{Z}_i$. Since the nonlinear parameters are specified, −2 ln likelihood is calculated as if $\mathbf{U}$ and $\mathbf{W}_i$ are known. First, the matrix product $\mathbf{M}_i = \mathbf{U}\mathbf{Z}_i'$ is calculated. $\mathbf{U}$ is upper triangular so some multiplications can be saved. In subscript notation, suppressing the subscript i which denotes the subject, this multiplication is

$$M_{\nu j} = \sum_{k=\nu}^{s} U_{\nu k} Z_{jk}. \tag{2.14}$$

There are two points to note about this equation. First, the lower limit of summation is ν rather than 1 since $\mathbf{U}$ is upper triangular.

Second, in matrix multiplication, the summation is usually over adjoining subscripts as in

$$\sum_{k=1}^{g} A_{\nu k} B_{kj},$$

which gives the νjth element of the matrix product. In equation (2.14), reversing the subscript on Z from kj to jk multiplies $\mathbf{U}$ by $\mathbf{Z}_i$ transposed. This also makes the multiplication dimensionally compatible since the second dimension (number of columns) of $\mathbf{Z}_i$ is g.

The total covariance matrix with σ^2 factored out is calculated as

$$\mathbf{V}_i = \mathbf{Z}_i \mathbf{U}'\mathbf{U}\mathbf{Z}_i' + \mathbf{W}_i = \mathbf{M}_i{}'\mathbf{M}_i + \mathbf{W}_i. \tag{2.15}$$

This matrix is augmented by adding columns of the $\mathbf{X}_i$ matrix and the $\mathbf{y}_i$ vector giving the augmented matrix

$$[\ \mathbf{V}_i \quad \mathbf{X}_i \quad \mathbf{y}_i\ ].$$

Applying the Cholesky factorization to this augmented matrix as in Graybill (1976, section 7.2, p 232) replaces the augmented matrix by

$$[\ \mathbf{T}_i \quad \mathbf{D}_i \quad \mathbf{b}_i\ ],$$

where $\mathbf{T}_i$ is the upper triangular factorization of $\mathbf{V}_i$,

$$\mathbf{V}_i = \mathbf{T}_i'\mathbf{T}_i, \tag{2.16}$$

and

$$\mathbf{D}_i = (\mathbf{T}_i')^{-1}\mathbf{X}_i \quad \text{and} \quad \mathbf{b}_i = (\mathbf{T}_i')^{-1}\mathbf{y}_i.$$

Since the inverse of the product of two matrices is the product of the inverses in the reverse order,

$$\mathbf{V}_i^{-1} = \mathbf{T}_i^{-1}(\mathbf{T}_i')^{-1},$$

we have

$$\mathbf{D}_i'\mathbf{D}_i = \mathbf{X}_i'\mathbf{T}_i^{-1}(\mathbf{T}_i')^{-1}\mathbf{X}_i = \mathbf{X}_i'\mathbf{V}_i^{-1}\mathbf{X}_i,$$

$$\mathbf{D}_i'\mathbf{b}_i = \mathbf{X}_i'\mathbf{T}_i^{-1}(\mathbf{T}_i')^{-1}\mathbf{y}_i = \mathbf{X}_i'\mathbf{V}_i^{-1}\mathbf{y}_i,$$

and

$$\mathbf{b}_i'\mathbf{b}_i = \mathbf{y}_i'\mathbf{T}_i^{-1}(\mathbf{T}_i')^{-1}\mathbf{y}_i = \mathbf{y}_i'\mathbf{V}_i^{-1}\mathbf{y}_i.$$

The matrix $\mathbf{D}_i'\mathbf{D}_i$ augmented by $\mathbf{D}_i'\mathbf{b}_i$ is summed over subjects producing the matrix

$$[\ \textstyle\sum_i \mathbf{D}_i'\mathbf{D}_i \quad \sum_i \mathbf{D}_i'\mathbf{b}_i\ ] = [\ \sum_i \mathbf{X}_i'\mathbf{V}_i^{-1}\mathbf{X}_i \quad \sum_i \mathbf{X}_i'\mathbf{V}_i^{-1}\mathbf{y}_i\ ], \tag{2.17}$$

which has dimension $b \times (b+1)$. This matrix contains the left hand matrix and the right hand vector of the normal equations (2.10) which are needed to calculate the estimated fixed effect regression coefficients $\hat{\boldsymbol{\beta}}$. However, to calculate −2 ln likelihood it is not necessary to actually solve for $\hat{\boldsymbol{\beta}}$. Applying the Cholesky factorization to this augmented matrix produces the matrix

$$[\ \mathbf{G} \quad \mathbf{r}\ ], \tag{2.18}$$

where $\mathbf{G}$ is upper triangular,

$$\mathbf{G}'\mathbf{G} = \sum_i \mathbf{X}_i'\mathbf{V}_i^{-1}\mathbf{X}_i,$$

and

$$\mathbf{r} = (\mathbf{G}')^{-1}(\sum_i \mathbf{X}_i'\mathbf{V}_i^{-1}\mathbf{y}_i).$$

Now

$$\mathbf{r}'\mathbf{r} = (\sum_i \mathbf{X}_i'\mathbf{V}_i^{-1}\mathbf{y}_i)'(\sum_i \mathbf{X}_i'\mathbf{V}_i^{-1}\mathbf{X}_i)^{-1}(\sum_i \mathbf{X}_i'\mathbf{V}_i^{-1}\mathbf{y}_i)$$

is the weighted regression sum of squares. The weighted total sum of squares (TSS) for the fixed effects regression is $\mathbf{b}_i'\mathbf{b}_i$ summed over subjects, so the mean square error is

$$\hat{\sigma}^2 = \frac{1}{n}\left(TSS - \sum_{k=1}^{b} r_k^2\right), \tag{2.19}$$

where r_k are the elements of $\mathbf{r}$. Because of the subtraction in this equation, the above calculations should be carried out in double precision. The weighted total sum of squares and the weighted regression sum of squares may be of the same order of magnitude so the subtraction causes the loss of significant digits. On many computers single precision is about 7 significant digits and double precision is about 14 significant digits.

The Cholesky factorization gives not only a simple and numerically stable method for calculating the quantities necessary for the weighted regression, but also the determinant which is necessary for -2 ln likelihood. The determinant of a triangular matrix is the product of the diagonal elements. The determinant of the product of two matrices is the product of the determinants. It follows that the determinant term in the likelihood (2.9) is calculated as the product of the squared diagonal elements of $\mathbf{T}_i$, therefore,

$$\ln|\mathbf{V}_i| = \sum_{j=1}^{n_i} \ln T_{jj}^2,$$

where T_{jj} is the jth diagonal element of $\mathbf{T}_i$. This completes the calculation of -2 ln likelihood, equation (2.9).

After the optimization is complete, the estimated regression coefficients can be calculated from $\mathbf{G}$ and $\mathbf{r}$,

$$\begin{aligned}\hat{\boldsymbol{\beta}} &= (\mathbf{G}'\mathbf{G})^{-1}(\sum_i \mathbf{X}_i'\mathbf{V}_i^{-1}\mathbf{y}_i)\\ &= \mathbf{G}^{-1}(\mathbf{G}')^{-1}(\sum_i \mathbf{X}_i'\mathbf{V}_i^{-1}\mathbf{y}_i)\\ &= \mathbf{G}^{-1}\mathbf{r},\end{aligned}$$

therefore, $\hat{\boldsymbol{\beta}}$ is the solution of the system of equations

$$\mathbf{G}\hat{\boldsymbol{\beta}} = \mathbf{r}.$$

Since $\mathbf{G}$ is upper triangular, this system of equations is solved easily by back substitution.

The estimated covariance matrix of $\hat{\boldsymbol{\beta}}$ is

$$\text{cov}(\hat{\boldsymbol{\beta}}) = \hat{\sigma}^2(\sum_i \mathbf{X}_i'\mathbf{V}_i^{-1}\mathbf{X}_i)^{-1}. \tag{2.20}$$

This inverse can be calculated directly from $\mathbf{G}$. FORTRAN subroutines for these calculations are given in the Appendix.

2.3 Variance heterogeneity

In biological data, it is common for the variance of the observations to increase as the magnitude of the numbers increases. In some situations, a log or square root transformation can be used to stabilize the variance of the observations. Another possibility is to introduce a variance functions as in quasi-likelihood methods.

For observation j on subject i, let $\mathbf{x}_{ij}$ be the corresponding row of the $\mathbf{X}_i$ matrix and $\mathbf{z}_{ij}$ be the corresponding row of the $\mathbf{Z}_i$ matrix. The subject's mean level at that observation is $\mathbf{x}_{ij}\boldsymbol{\beta}+\mathbf{z}_{ij}\boldsymbol{\gamma}_i$. Estimates of $\boldsymbol{\beta}$ and $\boldsymbol{\gamma}_i$ are available from the previous iteration in the nonlinear optimization process. Now set

$$\text{var}(\epsilon_{ij}) = \sigma^2(\mathbf{x}_{ij}\hat{\boldsymbol{\beta}} + \mathbf{z}_{ij}\hat{\boldsymbol{\gamma}}_i)^{power},$$

where *power* is non-negative. *power* $= 1$ replaces a possible square root transformation (the variance is proportional to the mean) while *power* $= 2$ replaces a possible log transformation (the standard deviation is proportional to the mean) for data with a constant coefficient of variation. The Gaussian -2 ln likelihood is still minimized since it contains the determinant term to account for changing the variance. It is possible to try various powers and choose the one that gives the lowest value of -2 ln likelihood.

The use of a variance function with Gaussian data does have some problems. Gaussian data can go negative, and the use of these variance functions requires that the estimated mean function be positive. In many applications this is no problem. When all the observations are positive, the mean functions tend to be positive. In certain nonlinear applications (see Chapter 7) it is possible to constrain the function to be positive. Care is also necessary if the mean function gets close to zero since these observations will be given large weight. With these warnings, the use of variance functions can be quite useful when all the observations are clearly above zero.

2.4 Residuals

A method of checking the calculations and avoiding the subtraction in (2.19) is to calculate $\hat{\boldsymbol{\beta}}$ and make another pass through the data calculating residuals and the weighted residual sum of squares. This costs computer time but produces a more stable numerical method of calculating the weighted residual sum of squares. The residual vector for each subject is

$$\mathbf{e}_i = (\mathbf{y}_i - \mathbf{X}_i\hat{\boldsymbol{\beta}}),$$

and the weighted residual sum of squares for this subject is

$$(\mathbf{y}_i - \mathbf{X}_i\hat{\boldsymbol{\beta}})'\mathbf{V}_i^{-1}(\mathbf{y}_i - \mathbf{X}_i\hat{\boldsymbol{\beta}}) = \mathbf{e}_i'\mathbf{V}_i^{-1}\mathbf{e}_i.$$

Calculating $\mathbf{V}_i$ from equation (2.15), and augmenting it by the vector of residuals gives the augmented matrix

$$[\ \mathbf{V}_i \quad \mathbf{e}_i\ ] .$$

Applying the Cholesky factorization to this augmented matrix replaces the augmented matrix by

$$[\ \mathbf{T}_i \quad \tilde{\mathbf{e}}_i\ ] ,$$

where

$$\tilde{\mathbf{e}}_i = (\mathbf{T}_i')^{-1}\mathbf{e}_i, \tag{2.21}$$

and $\mathbf{T}_i$ is the upper triangular factor of $\mathbf{V}_i$ in equation (2.16). The weighted sum of squares of residuals for this subject is the sum of squares of the elements of the vector $\tilde{\mathbf{e}}_i$. Summing over all the subjects and dividing by n gives the alternate way of calculating $\hat{\sigma}^2$.

The vector of transformed residuals, (2.21), are natural residuals to examine for outliers, serial correlation, normality, etc. Premultplying the errors, ϵ_i, in (2.1) by $(\mathbf{T}_i')^{-1}$ produces transformed errors with covariance matrix $\sigma^2\mathbf{I}$. Since $\mathbf{T}_i$ is upper triangular, $\mathbf{T}_i'$ is lower triangular, as is $(\mathbf{T}_i')^{-1}$. Premultplying the error vector by this lower triangular matrix produces transformed errors that are uncorrelated with constant variance. Each transformed error is a linear combination of that error and all the previous errors. The linear combinations are the particular linear combinations that make each transformed error orthogonal to the previous transformed errors. This procedure could be called recursive orthogonalization, and is carried out using the Cholesky decomposition. When the lower triangular matrix $(\mathbf{T}_i')^{-1}$ is applied to the residuals, the resulting transformed residuals will be approximately uncorrelated with constant variance.

2.5 Restricted maximum likelihood

Restricted maximum likelihood (REML), also known as residual maximum likelihood was introduced on page 20 for a balanced repeated measures design. Laird and Ware (1982) present a unified discussion of restricted maximum likelihood estimation from a Bayesian point of view. The purpose of restricted maximum likelihood estimation is to give less biased estimates of the variances than maximum likelihood estimates. This is most important when the number of fixed parameters is not small relative to the total number of observations.

For given values of the $\mathbf{B}$ and $\mathbf{W}_i$ matrices, the REML estimate of $\boldsymbol{\beta}$ is the same as equation (2.11) for maximum likelihood estimation (Diggle, 1988),

$$\hat{\boldsymbol{\beta}} = (\sum_i \mathbf{X}_i'\mathbf{V}_i^{-1}\mathbf{X}_i)^{-1}(\sum_i \mathbf{X}_i'\mathbf{V}_i^{-1}\mathbf{y}_i).$$

However, the unbiased estimate of σ^2 is used,

$$\hat{\sigma}^2 = \frac{1}{n-b}\sum_i (\mathbf{y}_i - \mathbf{X}_i\boldsymbol{\beta})'\mathbf{V}_i^{-1}(\mathbf{y}_i - \mathbf{X}_i\boldsymbol{\beta}), \tag{2.22}$$

where b is the number of fixed regression coefficients, the length of $\boldsymbol{\beta}$. The function to be minimized with respect to the parameters of the covariance matrix is

$$\begin{aligned} \ell^* &= n\ln(2\pi) + (n-b)\ln\hat{\sigma}^2 \\ &+ \ln|\sum_i \mathbf{X}_i'\mathbf{V}_i^{-1}\mathbf{X}_i| + \sum_i \ln|\mathbf{V}_i| + n - b. \end{aligned} \tag{2.23}$$

Hocking (1985, Chapter 8) gives a derivation of REML estimates and refers to ℓ^* in (2.23) as a modified -2 ln likelihood. Hocking also suggests that modified likelihood ratio tests can be used for hypothesis testing.

2.6 The EM algorithm

Laird and Ware (1982) use an interesting variation of the EM algorithm (for Expectation, Maximization) to estimate parameters in equation (2.1). The EM algorithm is usually used with data that have missing observations. Here the 'missing observations' are taken to be the unobservable random effects and errors, $\boldsymbol{\gamma}_i$ and $\boldsymbol{\epsilon}_i$. Two variations of the algorithm can be used to obtain maximum likelihood estimates or REML estimates of $\boldsymbol{\beta}$, $\boldsymbol{\gamma}_i$ and $\boldsymbol{\epsilon}_i$.

G. K. Robinson (1991) gives an excellent discussion of best linear unbiased prediction or BLUP as a method of estimating random effects. Classical statisticians usually do not estimate random effects while Bayesian statisticians do this routinely. The estimation of random effects is quite useful for looking at individual subjects' deviation from their group mean, and in the calculation of residuals.

The equations for obtaining maximum likelihood estimates using the EM algorithm follow for the case when the within subject errors are uncorrelated, $\mathbf{W}_i = \mathbf{I}$, so that

$$\mathbf{V}_i = \mathbf{Z}_i\mathbf{B}\mathbf{Z}_i' + \mathbf{I}.$$

For given initial values of $\mathbf{B}$, $\boldsymbol{\beta}$ is estimated using equation (2.11), and the random effects for each subject are estimated (or predicted) as

$$\hat{\boldsymbol{\gamma}}_i = \mathbf{B}\mathbf{Z}_i'\mathbf{V}_i^{-1}(\mathbf{y}_i - \mathbf{X}_i\hat{\boldsymbol{\beta}}). \tag{2.24}$$

The within subject error vector, $\boldsymbol{\epsilon}_i$, is estimated using residuals,

$$\mathbf{e}_i = \mathbf{V}_i^{-1}(\mathbf{y}_i - \mathbf{X}_i\hat{\boldsymbol{\beta}}). \tag{2.25}$$

Now the scalar, t_1 and the matrix $\mathbf{t}_2$ are calculated,

$$t_1 = \sum_i \{\mathbf{e}_i'\mathbf{e}_i + \hat{\sigma}^2\text{trace}(\mathbf{I} - \mathbf{V}_i^{-1})\}, \tag{2.26}$$

$$\mathbf{t}_2 = \sum_i \{\hat{\boldsymbol{\gamma}}_i\hat{\boldsymbol{\gamma}}_i' + \hat{\sigma}^2(\mathbf{B} - \mathbf{B}\mathbf{Z}_i'\mathbf{V}_i^{-1}\mathbf{Z}_i\mathbf{B})\}, \tag{2.27}$$

where the trace of a matrix is the sum of the diagonal elements. Equations (2.26) and (2.27) define the E step of the EM algorithm. The M step consists of the following two equations,

$$\hat{\sigma}^2 = t_1/n, \tag{2.28}$$

and

$$\hat{\mathbf{B}} = \mathbf{t}_2/(m\hat{\sigma}^2), \tag{2.29}$$

where m is the number of subjects. This procedure is iterated until the estimates stabilize.

The principle behind the EM algorithm is that if the within subject errors, $\boldsymbol{\epsilon}_i$, and the random effects, $\boldsymbol{\gamma}_i$, could actually be observed, $\sum_i \boldsymbol{\epsilon}_i'\boldsymbol{\epsilon}_i$ and $\sum_i \boldsymbol{\gamma}_i\boldsymbol{\gamma}_i'$ would be 'sufficient statistics' for estimating σ^2 and $\mathbf{B}$. Since $\boldsymbol{\epsilon}_i$ and $\boldsymbol{\gamma}_i$ are not observed, their estimates, $\mathbf{e}_i$ and $\hat{\boldsymbol{\gamma}}_i$, with bias corrections, are used to estimate the sufficient statistics t_1 and $\mathbf{t}_2$. This gives t_1 and $\mathbf{t}_2$ the proper expectation in the E step of the EM algorithm.

Maximum likelihood estimates of the variance components fail to take into account the degrees of freedom lost by estimating the fixed effects, $\boldsymbol{\beta}$, and are biased towards zero. The EM algorithm can also be used to obtain REML estimates which correct for this bias (Laird, 1982). The modifications are in equations (2.26) and (2.27).

$$t_1^* = t_1 + \hat{\sigma}^2 \sum_i \text{trace}\{\mathbf{V}_i^{-1}\mathbf{X}_i(\sum_k \mathbf{X}_k'\mathbf{V}_k^{-1}\mathbf{X}_k)^{-1}\mathbf{X}_i'\mathbf{V}_i^{-1}\}. \quad (2.30)$$

and,

$$\mathbf{t}_2^* = \mathbf{t}_2 + \hat{\sigma}^2\mathbf{B}[\sum_i \{\mathbf{Z}_i'\mathbf{V}_i^{-1}\mathbf{X}_i(\sum_k \mathbf{X}_k'\mathbf{V}_k^{-1}\mathbf{X}_k)^{-1}\mathbf{X}_i'\mathbf{V}_i^{-1}\mathbf{Z}_i\}]\mathbf{B} \quad (2.31)$$

The sufficient statistics for the REML estimates, t_1^* and $\mathbf{t}_2^*$, are the sufficient statistics for the maximum likelihood estimates, t_1 and $\mathbf{t}_2$, plus non-negative terms, which correct for the downward bias.

One problem with the EM algorithm is that it has relatively slow (linear) convergence compared to quasi-Newton methods of nonlinear optimization (Dennis and Schnabel, 1983). With a version of the quasi-Newton method that does not require derivatives of the function being minimized, a subroutine to evaluate −2 ln likelihood, or the modified −2 ln likelihood in the REML case, is all that is necessary. A second problem with the EM algorithm is that it does not generalize easily to the case of serially correlated errors discussed in the next chapter. For an autoregressive within subject error structure with unequally spaced observations, even if the within subject errors could be observed, the only sufficient statistic for the autoregressive parameter is the complete set of observations. This means that the EM algorithm must have a second level of iterations within the M step of the algorithm. Iterations within iterations are best avoided by finding a single objective function to be minimized using a nonlinear optimization routine.

2.7 Predictions

After fitting a model, it is possible to calculate predicted values of the mean curves with estimated standard errors. With mixed models, some interesting new concepts arise when making predictions for individual subjects.

2.7.1 *Population mean curves*

The prediction of population mean curves is the same as in ordinary regression. For each subject in the study, there is an $\mathbf{X}_i$ matrix. Let $\mathbf{x}$ denote a possible row of $\mathbf{X}_i$ for any subject. For example, if $\mathbf{X}_i$ consists of two columns for fitting a straight line in time,

$$\mathbf{x} = [1 \; t],$$

where t denotes the time at which the population mean line is to be estimated. It may or may not be an actual time used in the study. If $\mathbf{X}_i$ consists of three columns, the third being a (0,1) indicator variable denoting one of two groups,

$$\mathbf{x} = [1 \; t \; 0] \quad \text{or} \quad \mathbf{x} = [1 \; t \; 1]$$

depending on for which group the line is to be estimated. The estimated population mean for a given $\mathbf{x}$ vector is

$$\hat{y} = \mathbf{x}\hat{\boldsymbol{\beta}}, \tag{2.32}$$

and has estimated variance

$$\text{var}(\hat{y}) = \mathbf{x}\,\text{cov}(\hat{\boldsymbol{\beta}})\mathbf{x}' = \hat{\sigma}^2\mathbf{x}(\sum_i \mathbf{X}_i'\mathbf{V}_i^{-1}\mathbf{X}_i)^{-1}\mathbf{x}'. \tag{2.33}$$

The estimated standard error of the estimate is the square root of this estimated variance. By varying the elements of the $\mathbf{x}$ vector, estimated population curves can be generated for different groups or different values of covariates.

2.7.2 *Individual predictions*

First, consider the prediction of a new subject for which there have been no data collected. Let $\mathbf{x}$ represent a possible row of an $\mathbf{X}_i$ matrix for this subject, and $\mathbf{z}$ represent a possible row of the corresponding $\mathbf{Z}_i$ matrix. Since the prior estimates of the random effects, $\boldsymbol{\gamma}_i$, are zero, the prediction is the same as for the mean line (2.32), but the estimated variance, from the model, (2.1), is

$$\mathbf{x}\,\text{cov}(\hat{\boldsymbol{\beta}})\mathbf{x}' + \hat{\sigma}^2\mathbf{z}\mathbf{B}\mathbf{z}' + \hat{\sigma}^2. \tag{2.34}$$

The second and third terms represent the contributions from the between subject and within subject components of variance.

Now, consider the prediction for a subject who has previous observations. This means that there is information about the subject's random effects that can be used in the prediction, and this reduces the variance. The prediction of a subject's random effects is given by (2.24), and the estimated covariance matrix of this prediction is (Laird and Ware, 1982),

$$\begin{aligned} \mathrm{cov}(\hat{\gamma}_i - \gamma_i) &= \hat{\sigma}^2\{\mathbf{B} - \mathbf{B}\mathbf{Z}_i'\mathbf{V}_i^{-1}\mathbf{Z}_i\mathbf{B} \\ &+ \mathbf{B}\mathbf{Z}_i'\mathbf{V}_i^{-1}\mathbf{X}_i(\sum_k \mathbf{X}_k'\mathbf{V}_k^{-1}\mathbf{X}_k)^{-1}\mathbf{X}_i'\mathbf{V}_i^{-1}\mathbf{Z}_i\mathbf{B}\}. \end{aligned} \tag{2.35}$$

If the within subject covariance matrix $\sigma^2\mathbf{W}_i = \sigma^2\mathbf{I}$, then the within subject errors are uncorrelated and there is no information in the previous within subject errors about the within subject error at the point of the prediction. The estimated prediction for this subject is

$$\hat{y} = \mathbf{x}\hat{\beta} + \mathbf{z}\hat{\gamma}_i,$$

and the estimated prediction variance is

$$\mathbf{x}\,\mathrm{cov}(\hat{\beta})\mathbf{x}' + \mathbf{z}\mathrm{cov}(\hat{\gamma}_i - \gamma_i)\mathbf{z}' + \hat{\sigma}^2, \tag{2.36}$$

which is smaller than (2.34).

2.8 Model selection

When two models are fit to the same data by maximum likelihood, and one model is a constrained version of the other, the likelihood ratio test can be used to test the null hypothesis that the model with more parameters is not a significantly better fit than the model with fewer parameters. The statement that one model is a constrained version of the other usually means that the second model has all the parameters of the first model plus a few extra. In this case the constraints in the reduced model refer to the fact that some of the parameters have been constrained to be zero. However, other versions of constrained models are possible; e.g. a model can be reduced by constraining two parameters to be the same. In any event, the likelihood ratio test is not an exact test but an asymptotic test which assumes that the sample size is large. What is meant by large is not always clear, but likelihood ratio tests often give reasonable results for moderate sample sizes, say twenty subjects with six observations per subject. In longitudinal data analysis problems, there is also the question of which sample size should approach infinity for asymptotic results to hold. It is possible to have a relatively small number of subjects with a large number of observations per subject, or a large number of subjects with a small number of observations per subject. The first case provides substantial information about $\sigma^2\mathbf{W}_i$, the within subject component of variance, whereas the second case is more informative for $\sigma^2\mathbf{B}$, the between subject component of variance. Asymptotic results in time series analysis require that the number of observations per subject gets large; otherwise, estimates may not converge to the true values of the parameters. In multivariate analysis, the number of observations per subject is fixed and the number of subjects gets large. Here, we will take the point of view that our sample size is fixed, nothing is going to infinity and that the likelihood ratio test is an approximate test which helps in model selection.

The simplest application of the likelihood ratio test is to fit a model with a certain number of parameters. This includes both linear and nonlinear parameters. Second, a model is fit which has the same parameters plus r extra parameters. In the second model, the minimized value of -2 ln likelihood must be smaller than the minimized value of -2 ln likelihood in the first model since some constraints have been removed. In nonlinear optimization problems sometimes -2 ln likelihood increases when parameters are added to the model. In theory this cannot happen, but sometimes the program converges to a local minimum which is higher than the minimum of the first model. One way to prevent convergence to a different local minimum is to start the second model from values of the parameters which minimized the first model with the new parameters set equal to zero. This starts the optimization from the minimum of the previous optimization and a good algorithm will go only in a downhill direction. Because of the risk of ending up at a local minimum rather than at the global minimum, it is always good practice to try several starting values of the parameters to see if they converge to the same point.

If ℓ_1 is the value of -2 ln likelihood from the first model and ℓ_2 is the value from the second model with r extra parameters, the likelihood ratio test states that under the null hypothesis the two models are the same (i.e. the extra parameters do not improve the fit) and the difference in the values of the two -2 ln likelihoods is asymptotically distributed as chi-square with r degrees of freedom,

$$\ell_1 - \ell_2 \sim \chi^2_r.$$

A large value of this difference rejects the null hypothesis and accepts the alternative hypotheses that there is a significant improvement in the fit when the extra parameters are introduced.

When many models are fit to the same data, a likelihood ratio test can be calculated between any two models where one is a constrained version or reduced version of the other. Two such models are called 'nested'. This can get quite confusing and, when many models are fit, there are often pairs which are not nested so the likelihood ratio test can not be used to test these differences. An overall model selection procedure which is quite helpful is Akaike's Information Criterion (AIC) (Akaike, 1973a; 1974a; Sakamoto et al., 1986; Bozdogan, 1987). AIC is based on decision theory and penalizes −2 ln likelihood for the number of parameters fit to the data to avoid overfitting:

$$\text{AIC} = \ell + 2(\text{number of estimated parameters}).$$

For every model under consideration, ℓ is calculated and twice the number of parameters that were estimated is added to ℓ to give AIC for the model. The model which has the lowest value of AIC is selected as the 'best' model.

AIC has received some criticism in the time series analysis literature because it is not a consistent estimate of the order of an autoregression. That is to say, if an autoregression has true order p and the number of observations on the time series goes to infinity, AIC does not always select order p. Sometimes it chooses too large an order. Schwarz (1978) suggested modified AIC,

$$\text{SC} = \ell + (\ln n)(\text{number of estimated parameters}),$$

where n is the total number of observations on all subjects. SC has an increased penalty for overfitting compared to AIC. Bozdogan (1987) suggests increasing the penalty term slightly more, and called the result the consistent AIC,

$$\text{CAIC} = \ell + (1 + \ln n)(\text{number of estimated parameters}).$$

I prefer to use AIC with the slight variation that models that are within two units of the lowest AIC are considered to be competitive models for the best. From the competitive models, the one with the fewest parameters is usually selected. This version of AIC has some theoretical justification (Duong, 1984).

2.9 Testing contrasts

The fitting of two models by nonlinear optimization for the purpose of hypothesis testing using the likelihood ratio test can be avoided by fitting the more complete model (the model with more parameters) and testing contrasts on the β's. The simplest example of a contrast is the test of the hypothesis that one of the coefficients is equal to zero. The use of the likelihood ratio test would fit two models with and without the presence of this parameter and compare the change in -2 ln likelihood to chi-square with one degree for freedom. The hypothesis being tested is

$$H_0: \ \beta_k = 0,$$

where k denotes the parameter being tested. Another way to test this hypothesis is to use the estimated covariance matrix of the estimated β's shown in (2.20). The kth diagonal element of this matrix is the estimated variance of β_k. The test statistic, often referred to as Wald's test, is

$$t_{df} = \hat{\beta}_k / \sqrt{\text{cov}(\hat{\beta})_{kk}},$$

where $\text{cov}(\hat{\beta})_{kk}$ is the kth diagonal element of the covariance matrix, or the equivalent test statistic

$$F_{1,df} = \hat{\beta}_k^2 / \text{cov}(\hat{\beta})_{kk}.$$

In linear regression (Draper and Smith, 1981), this would be an exact t-test or F-test for the hypothesis assuming independent Gaussian errors with constant variance. Only in the case of testing a single coefficient can the t-statistic be squared to give an F-statistic with one degree of freedom in the numerator. When testing more than one coefficient simultaneously, the F-test is the appropriate test.

When using nonlinear estimation, this test is no longer exact, and there is some question as to whether it should be called a t-test or an F-test, and what degrees of freedom (df) should be used. Wald's test is asymptotically equivalent to the likelihood ratio test for large sample sizes which means that the degrees of freedom are large. In this case,

$$t_{df} \sim z,$$

where z is a standard normal variable, and

$$F_{1,df} \sim \chi_1^2,$$

where χ_1^2 is chi-square with one degree of freedom. The symbol $\sim$ means 'approaches as the degrees of freedom becomes large'. These results can easily be verified by checking tables of the t distribution and the normal distribution, and tables of the F distribution and the chi-square distribution.

For moderate degrees of freedom, the case for using the t-test or F-test rather than the limiting normal or chi-square test is that the t and F-tests are more conservative giving higher cut-off values for significance. The degrees of freedom are the total sample size for all subjects minus b, the number of elements of $\boldsymbol{\beta}$.

Consider now the two group experiment shown in equation (2.2) with four observations per subject. It may be of interest to test the hypothesis

$$H_0 : \beta_1 = \beta_2. \tag{2.37}$$

This is a contrast that can be written in matrix form as

$$\begin{bmatrix} 1 & -1 \end{bmatrix} \begin{bmatrix} \beta_1 \\ \beta_2 \end{bmatrix} = 0.$$

Defining a contrast matrix

$$\mathbf{C} = \begin{bmatrix} 1 & -1 \end{bmatrix},$$

the hypothesis (2.37) can be written

$$\mathbf{C}\boldsymbol{\beta} = 0.$$

This is a linear combination, and its variance depends on the variance of both coefficients and the covariance between them,

$$\text{var}(\mathbf{C}\hat{\boldsymbol{\beta}}) = \text{var}(\hat{\beta}_1) + \text{var}(\hat{\beta}_2) - 2\text{cov}(\hat{\beta}_1, \hat{\beta}_2).$$

If $\text{cov}(\hat{\boldsymbol{\beta}})$ is a 2×2 matrix, this can be written in matrix notation as

$$\text{var}(\mathbf{C}\hat{\boldsymbol{\beta}}) = \mathbf{C}\{\text{cov}(\hat{\boldsymbol{\beta}})\}\mathbf{C}'.$$

The test statistic is now

$$F_{1,df} = (\mathbf{C}\hat{\boldsymbol{\beta}})^2/\text{var}(\mathbf{C}\hat{\boldsymbol{\beta}})$$

which, again, is asymptotically distributed as chi-square with one degree of freedom.

Suppose there are three groups of subjects, and it is of interest to test whether all three coefficients are the same. This hypothesis is

$$H_0 : \beta_1 = \beta_2 \quad \text{and} \quad \beta_2 = \beta_3.$$

In matrix form, this is

$$\begin{bmatrix} 1 & -1 & 0 \\ 0 & 1 & -1 \end{bmatrix} \begin{bmatrix} \beta_1 \\ \beta_2 \\ \beta_3 \end{bmatrix} = \begin{bmatrix} 0 \\ 0 \end{bmatrix}.$$

In this case, the contrast matrix is

$$\mathbf{C} = \begin{bmatrix} 1 & -1 & 0 \\ 0 & 1 & -1 \end{bmatrix}.$$

There are now two contrasts to be tested simultaneously. The two linear combinations of the $\boldsymbol{\beta}$'s, $\mathbf{C}\boldsymbol{\beta}$, have a 2×2 covariance matrix

$$\text{cov}(\mathbf{C}\hat{\boldsymbol{\beta}}) = \mathbf{C}\{\text{cov}(\hat{\boldsymbol{\beta}})\}\mathbf{C}',$$

and the test statistic is

$$(\mathbf{C}\hat{\boldsymbol{\beta}})'[\mathbf{C}\{\text{cov}(\hat{\boldsymbol{\beta}}\})\mathbf{C}']^{-1}\mathbf{C}\hat{\boldsymbol{\beta}},$$

which under the null hypothesis is asymptotically distributed as chi-square with 2 degrees of freedom. This follows from a general result that if $\mathbf{x}$ is a Gaussian random vector with zero mean and covariance matrix $\mathbf{V}$, the quadratic form $\mathbf{x}'\mathbf{V}^{-1}\mathbf{x}$ has a chi-square distribution with degrees of freedom equal to the length of the vector $\mathbf{x}$ (see Exercise 3). The Cholesky factorization can be used to efficiently calculate the test statistic by augmenting the matrix $\mathbf{C}\{\text{cov}(\hat{\boldsymbol{\beta}}\})\mathbf{C}'$ by the vector $\mathbf{C}\hat{\boldsymbol{\beta}}$. After the factorization, sum the squares of the elements of the augmented vector as was done in (2.18) to calculate the weighted regression sum of squares.

The chi-square test statistic can be divided by its degrees of freedom to obtain an approximate F statistic as in the first example in this section.

2.10 An example

In this section, the analysis of the example in Chapter 1 will be given in more detail. The data shown in Table 1.1 have two groups with five subjects in each group, seven observations per subject with a single missing observation. The first model fit to the data is the saturated model that allows each group and each time point to have a different value. There are 14 linear parameters. This model has an arbitrary time effect, a group effect (which allows one group to be shifted vertically with respect to the other) and a group by time interaction which allows the two curves for the two groups to have different shapes. The saturated model written in partitioned matrix form is, for a subject in group 1

$$\mathbf{y}_{i(1)} = \begin{bmatrix} \mathbf{I} & \mathbf{0} \end{bmatrix} \begin{bmatrix} \boldsymbol{\beta}_1 \\ \boldsymbol{\beta}_2 \end{bmatrix} + \mathbf{1}\gamma_{i(1)} + \boldsymbol{\epsilon}_{i(1)},$$

and for a subject in group 2,

$$\mathbf{y}_{i(2)} = \begin{bmatrix} \mathbf{I} & \mathbf{I} \end{bmatrix} \begin{bmatrix} \boldsymbol{\beta}_1 \\ \boldsymbol{\beta}_2 \end{bmatrix} + \mathbf{1}\gamma_{i(2)} + \boldsymbol{\epsilon}_{i(2)}.$$

In these equations, $\mathbf{I}$ is a 7×7 identity matrix, $\mathbf{0}$ is a 7×7 matrix of zeros, the two $\boldsymbol{\beta}$'s are 7×1 column vectors, $\mathbf{1}$ is a 7×1 column vector of ones, $\gamma_{i(k)}$ is a scalar representing the random subject effect for subject $i(k)$ with unknown variance $\sigma^2 B$, and $\boldsymbol{\epsilon}_{i(k)}$ is a 7×1 column vector of within subject errors assumed to be uncorrelated with variance σ^2. $\boldsymbol{\beta}_1$ is the mean time response curve for group 1, and $\boldsymbol{\beta}_1 + \boldsymbol{\beta}_2$ is the time response curve for group 2. First order autoregressive, AR(1), within subject error structures, to be discussed later, were tried but no significant improvement in the fit was found. This model has a minimized value of $\ell = 575.55$, and a value of AIC=607.55. This increase of 32 accounts for the 14 linear parameters, the one nonlinear parameter B, and σ^2.

The second, reduced model fit to the data allows the response curve to have an arbitrary shape with one group shifted vertically with respect to the other. This is an arbitrary time effect, a group effect but no group by time interaction. This model for a subject in group 1 is

$$\mathbf{y}_{i(1)} = \begin{bmatrix} \mathbf{I} & \mathbf{0} \end{bmatrix} \begin{bmatrix} \boldsymbol{\beta}_1 \\ \beta_2 \end{bmatrix} + \mathbf{1}\gamma_{i(1)} + \boldsymbol{\epsilon}_{i(1)},$$

and, for a subject in group 2,

$$\mathbf{y}_{i(2)} = \begin{bmatrix} \mathbf{I} & \mathbf{1} \end{bmatrix} \begin{bmatrix} \boldsymbol{\beta}_1 \\ \beta_2 \end{bmatrix} + \mathbf{1}\gamma_{i(2)} + \boldsymbol{\epsilon}_{i(2)}. \tag{2.38}$$

The reduced model has 8 linear parameters. β_2 is a scalar representing the group effect and $\mathbf{0}$ is a column of zeros. The reduced model has a minimized value of $\ell = 583.67$, and a value of AIC=603.67. A likelihood ratio test for group by time interaction is obtained by taking the difference of the ℓ's for the two models, noting that ℓ must be smaller for the model with more parameters. This difference is $583.67 - 575.55 = 8.12$ and when tested as chi-square with 6 degrees of freedom (the difference in the number of parameters in the two models) is found to be not significant ($p > 0.2$). The choice of model two based on the likelihood ratio agrees with the choice based on AIC since the value is smaller for model 2. The results of fitting model 2 to these data are shown in Figure 1.4

2.11 Exercises

1. Give the form of $\mathbf{Z}_i$, $\mathbf{B}$ and $\mathbf{W}_i$ in equation (2.1) for compound symmetry.

2. Find some values of the ρ's in the covariance matrix (2.6) where the matrix is not positive definite.

3. Show that if $\mathbf{x}$ is a Gaussian random vector with zero mean and covariance matrix $\mathbf{V}$ which is of full rank, the quadratic form $\mathbf{x}'\mathbf{V}^{-1}\mathbf{x}$ has a chi-square distribution with degrees of freedom equal to the dimension of $\mathbf{x}$. Hint: Consider the Cholesky factorization, $\mathbf{V} = \mathbf{U}'\mathbf{U}$, where $\mathbf{U}$ is upper triangular. Form a new random vector $\mathbf{y} = (\mathbf{U}')^{-1}\mathbf{x}$. What is the covariance matrix of $\mathbf{y}$?

4. In the example in section 2.10, what are the equations of the reduced version of equations (2.38) that would be used to test whether there is a significant time effect? The chi-square test based on the change in -2 ln likelihood for these two models has how many degrees of freedom?

CHAPTER 3

First Order Autoregressive Errors

3.1 Introduction

When data are collected over time for each subject, serial correlation is often present. When data are serially correlated, observations that are closer together tend to have higher correlations than observations which are farther apart. One implication of serially correlated data is that if an observation is above the subject's mean level, the next observation taken a short time later will also tend to be above the subject's mean level. A set of serially correlated observations is often referred to as a time series. In fact, any set of observations collected over time is a time series. Most of the studies of time series analysis have concentrated on the analysis of a single long series of observations taken at equally space time points (Box and Jenkins, 1976). The purpose of a time series analysis may be to determine the correlation structure of the data in order to make forecasts.

In longitudinal data analysis, the usual data structure is a number of short time series. Each time series consists of the observations taken on one subject, and the observations may or may not be equally spaced. The purpose of the analysis is to determine and model the correlation structure of the data so that the statistical estimates have known properties. For example, in regression analysis, the usual assumption is that the errors are uncorrelated with constant variance and have a Gaussian distribution. If these assumptions hold, and the regression model is correct, the usual t-tests, F-tests and confidence intervals for regression coefficients based on the t-statistic are correct. If the errors have positive serial correlation, confidence intervals based on the t-statistic will be too narrow, and hypothesis tests will be more likely to produce falsely significant results. If confidence intervals for the true mean were then constructed by assuming that the data were uncorrelated, the intervals would be too narrow. In repeated simulations,

95% confidence intervals would not capture the true mean 95% of the time. This problem is remedied by including serial correlation in the model for the within subject error structure.

In time series analysis, the word *realization* is used to describe independent time series with the same correlation structure. In longitudinal data analysis, different subjects are usually assumed to generate data with errors that are independent realizations of the same correlation structure.

3.2 Equally spaced observations

In their paper on longitudinal data analysis, Potthoff and Roy (1964) used an AR(1) error structure for equally spaced observations. AR(1) error structure is the within subject error structure shown in equation (2.5). Equally spaced observations include the case of equally spaced data with missing observations. There is a basic sampling interval.

An AR(1) time series with zero mean is generated by the equation

$$\epsilon_j = \phi\epsilon_{j-1} + \eta_j, \tag{3.1}$$

where η_j is a sequence of uncorrelated identically distributed random variables with zero mean and variance σ_η^2 that are also uncorrelated with the past of the ϵ_j process. When calculating likelihoods, the η_j's are assumed to have Gaussian distributions. The η_j's are also referred to as the innovations or random shocks (Box and Jenkins, 1976) of the AR(1) process. The term autoregression comes from this model since the present value of ϵ_j is regressed on the previous value.

The concept of a stationary time series is that the process is in equilibrium. It started up sometime in the past and has reached a steady state. Its statistical properties depend only on time differences, not absolute time. For this to be the case, it is necessary and sufficient that

$$-1 < \phi < 1.$$

Intuitively, this allows the effects of the past to die away with time. If $\phi = 1$, this process is a random walk,

$$\epsilon_j = \epsilon_{j-1} + \eta_j. \tag{3.2}$$

Suppose a random walk starts at the value 0 at observation 0, $\epsilon_0 = 0$. The random walk at observation $j > 0$ is the sum of j uncorrelated random variables,

$$\epsilon_j = \eta_1 + \eta_2 + \cdots + \eta_j.$$

The variance increases proportional to time and the process is not stationary. A stationary process has a variance which is constant over time.

Usually ϕ is in the range $0 \leq \phi < 1$, but negative values of ϕ are possible. Negative values of ϕ occur when adjoining observations are negatively correlated. It is possible for adjoining observations to appear to be negatively correlated because a mean value function has not been properly modelled. For example, if two observations are taken on each subject each day at, say, 8 a.m. and 8 p.m., an apparent negative serial correlation can be caused by the subject's circadian rhythm. Many physiological variables have a cyclic variation with a period of 24 hours. If this cyclic variation is modelled as part of the mean function, the negative correlation in the deviations about the mean would probably disappear.

The covariance function of a stationary time series depends only on the time difference between two observations. For a zero mean process, the covariance function is defined as

$$C(k) = E(\epsilon_j \epsilon_{j-k}),$$

where E denotes the expected or mean value, and k is the time difference between the two observations, the time 'lag'. The covariance function is symmetric about zero,

$$C(-k) = C(k). \tag{3.3}$$

The covariance function at lag zero is the variance of the process,

$$E(\epsilon_j^2) = C(0).$$

If equation (3.1) is multiplied by a past value of the process, ϵ_{j-k} where $k > 0$, and expected values taken, the following equation results,

$$C(k) = \phi C(k-1) \quad \text{for} \quad k > 0, \tag{3.4}$$

since

$$E(\epsilon_{j-k}\eta_j) = 0 \quad \text{for} \quad k > 0.$$

Equation (3.4) is known as the Yule-Walker equation for an AR(1) process, and shows the exponential decay of the covariance function as a function of the time lag,

$$C(k) = \phi^k C(0).$$

The variance of an AR(1) process depends on ϕ and σ_η^2. If equation (3.1) is multiplied by ϵ_j and expectations taken,

$$E(\epsilon_j^2) = E[\epsilon_j(\phi\epsilon_{j-1} + \eta_j)].$$

On the right hand side of this equation, ϵ_j is correlated with η_j. This problem can be resolved by substitution from equation (3.1),

$$\begin{aligned} C(0) &= E[(\phi\epsilon_{j-1} + \eta_j)(\phi\epsilon_{j-1} + \eta_j)] \\ &= \phi^2 C(0) + \sigma_\eta^2, \end{aligned}$$

since η_j is uncorrelated with ϵ_{j-1}. Now,

$$C(0) = \frac{\sigma_\eta^2}{1 - \phi^2}.$$

This gives the relationship between the variance of the process,

$$\text{var}(\epsilon_j) = C(0),$$

and the variance of the random input η_j.

The correlation function of a stationary time series is the covariance function divided by the variance of the process, and is the correlation between two observations as a function of the time separation. To obtain a correlation matrix for the within subject errors as in equation (2.5), with ones on the diagonal, let

$$C(0) = \sigma^2 = \frac{\sigma_\eta^2}{1 - \phi^2}. \tag{3.5}$$

so that

$$\text{var}(\epsilon_j) = \sigma^2.$$

Suppose that a subject who should have four equally spaced observations has the third observation missing. The within subject error covariance structure can be obtained from equation (2.5) by deleting the row and column corresponding to the missing observation, giving

$$\mathbf{W}_i = \begin{bmatrix} 1 & \rho & \rho^3 \\ \rho & 1 & \rho^2 \\ \rho^3 & \rho^2 & 1 \end{bmatrix}.$$

This matrix no longer has the banded or Toeplitz structure. The methods of the previous chapter can then be applied when there is an AR(1) within subject error structure with missing observations. If random within subject error is included in the model with variance $\sigma^2\sigma_o^2$, this matrix is

$$\mathbf{W}_i = \begin{bmatrix} 1+\sigma_o^2 & \rho & \rho^3 \\ \rho & 1+\sigma_o^2 & \rho^2 \\ \rho^3 & \rho^2 & 1+\sigma_o^2 \end{bmatrix}.$$

3.3 Unequally spaced observations

When observations can be taken on subjects at arbitrary time points, there must be an underlying continuous time process (Jones, 1981, 1985b; Diggle, 1988, 1990; Chi and Reinsel, 1989). For equally spaced observations, there may or may not be an underlying continuous time process. Some processes are truly discrete. For example, the maximum temperature each day is a discrete time process with one observation per day. Unequally spaced observations differ from equally spaced observations with some missing observations in that there is no basic sampling interval. Sometimes there is a basic sampling interval, but it is so small that almost all of the observations would be missing. For example, if a subject is followed for many years with only a few observations each year taken at arbitrary times, the day of the observation will be recorded. These observations could be considered daily observations with most of the observations missing, or the underlying process could be considered a continuous time process with unequally spaced observations. The latter would be a better description of the process.

The equally spaced model (3.1) is a difference equation driven by 'white noise'. White noise is a term used in time series analysis for a sequence of independent random variables (Box and Jenkins, 1976). A model for a continuous time process is a differential equation driven by 'white noise'. This white noise is in quotation marks because there is a problem with the existence of white noise in continuous time. Continuous time white noise exists only in the sense that its integral is a continuous time random walk often referred to as a Brownian motion or a Wiener process. A continuous time random walk or Wiener process is the limit of a discrete time random walk (3.2) as the time interval gets small, i.e.

$$w(t + \delta t) = w(t) + \eta(t),$$

where the $\eta(t)$ are independent random variables with zero means and variances $G\delta t$. As in discrete time random walks (3.2), the variance of a continuous time random walk grows in proportion to the length of the time interval, and G is the constant of proportionality. In the limit as δt approaches zero, the variance of $\eta(t)$ also approaches zero. A Wiener process can be simulated on a computer by this method (see exercise 2). If δt is small enough, the path function will be similar to a particle undergoing Brownian motion. The path function will be continuous but will have a very ragged appearance and its derivative does not exist. In fact, it is not of bounded variation. If a finite segment of this curve is stretched out straight, it would be infinitely long. With all these bad properties, the Wiener process is still the key to getting random input into a continuous time process.

The mathematical model for a continuous time AR(1) process, denoted CAR(1), is (Jones and Boadi-Boateng, 1991),

$$\frac{d}{dt}\epsilon(t) + \alpha_0\epsilon(t) = G\eta(t),$$

where $\eta(t)$ is continuous time 'white noise', i.e.

$$\eta(t) = \frac{d}{dt}w(t),$$

the derivative of a Wiener process. Since this derivative does not exist, the proper way to write the model is in differential form,

$$d\epsilon(t) + \alpha_0\epsilon(t)dt = G\,dw(t). \tag{3.6}$$

Although $dw(t)$ has the same existence problems, when (3.6) is solved by integration, a proper solution is obtained. The Wiener process is assumed to have unit variance per unit time, and the constant G in front of $dw(t)$ scales the input so that it has variance G^2 per unit time.

Consider the solution to (3.6) with the random input removed,

$$\frac{d}{dt}\epsilon(t) + \alpha_0\epsilon(t) = 0. \tag{3.7}$$

If (3.7) is integrated from time t_1 to time t_2, the solution is a prediction,

$$\epsilon(t_2) = \exp\{-\alpha_0(t_2 - t_1)\}\epsilon(t_1). \tag{3.8}$$

The solution, (3.8), is now in the form of a discrete time AR(1) process with an autoregression coefficient

$$\phi(t_2 - t_1) = \exp\{-\alpha_0(t_2 - t_1)\} \tag{3.9}$$

that depends on the time interval. For the solution to be a stable or stationary process it is necessary that the influence of the value of the process at time t_1 dies away with time. As $t_2 - t_1$ gets large, $\phi(t_2 - t_1)$ must get small. This will be the case if and only if α_0 is positive. The condition for stationarity of a CAR(1) process is

$$\alpha_0 > 0,$$

which implies that

$$0 < \phi(t_2 - t_1) \leq 1.$$

$\phi(t_2 - t_1)$ can take the value 1 only when the time interval is zero.

The autoregression coefficient in equation (3.9) is also the correlation between two observations separated by a time interval of $t_2 - t_1$, and it depends on a single parameter, α_0. The within subject covariance matrix, $\mathbf{W}_i$, therefore is parameterized in terms of the parameter α_0. For three unequally spaced observations with random within subject error, $\mathbf{W}_i$ would be of the form

$$\begin{bmatrix} 1+\sigma_o^2 & \exp\{-\alpha_0(t_2-t_1)\} & \exp\{-\alpha_0(t_3-t_1)\} \\ \exp\{-\alpha_0(t_2-t_1)\} & 1+\sigma_o^2 & \exp\{-\alpha_0(t_3-t_2)\} \\ \exp\{-\alpha_0(t_3-t_1)\} & \exp\{-\alpha_0(t_3-t_2)\} & 1+\sigma_o^2 \end{bmatrix}.$$

For a given value of α_0, $\mathbf{W}_i$ can be used in the likelihood in section 2.2, and nonlinear optimization can be used to find the maximum likelihood estimate of α_0. A log transformation can be used to constrain the estimate of α_0 to be positive.

The effect of the random input on the process can be developed intuitively. While integrating the process from t_1 to t_2, small random inputs, $dw(t)$, enter at each time. As the influence of the initial condition dies away exponentially with time, so does the effect of the random inputs. At time t_2, the influence of the random input at some earlier time is downweighted by the factor $\exp\{-\alpha_0(t_2 - t)\}$. The result can be integrated over the time interval to determine the properties of the random input over a finite time interval,

$$\eta(t_2 - t_1) = \int_{t_1}^{t_2} \exp\{-\alpha_0(t_2 - t)\} G\, dw(t).$$

From the properties of a Wiener process, all the infinitesimally small increments, $dw(t)$, are independent with zero mean and variance dt.

To calculate the variance of $\eta(t_2-t_1)$, the variances of the small increments can be integrated over the time interval remembering that the variance of a constant multiplied by a random variable is the constant squared multiplied by the variance of the random variable. Let $Q(t_2 - t_1) = \text{var}\{\eta(t_2 - t_1)\}$, then

$$\begin{aligned} Q(t_2 - t_1) &= G \int_{t_1}^{t_2} \exp\{-2\alpha_0(t_2 - t)\} dt \\ &= \frac{G}{2\alpha_0}[1 - \exp\{-2\alpha_0(t_2 - t_1)\}]. \qquad (3.10) \end{aligned}$$

The two important equations are (3.8) and (3.10). Equation (3.8) is a prediction of a zero mean CAR(1) process over a time interval of $t_2 - t_1$, and equation (3.10) is the variance of that prediction. Consider the two extremes of predicting over very short and very long time intervals. Since α_0 must be positive, as the prediction interval approaches zero, the prediction approaches the initial value and the prediction variance approaches zero. A prediction over a very long time interval approaches zero, the process mean, and the prediction variance approaches $G/2\alpha_0$, the process variance.

If the CAR(1) process is sampled at equally spaced intervals with a spacing δt, a discrete time AR(1) process is obtained with autoregression coefficient

$$\phi(\delta t) = \exp(-\alpha_0 \delta t).$$

Both the discrete and CAR(1) processes have a scale constant related to the variance of the random input. Since the variance of the continuous time process and the sampled version must be the same, let σ^2 be the variance of the process. Then

$$\sigma^2 = G/2\alpha_0, \tag{3.11}$$

and the variance of the random input over a step of length δt is

$$\begin{aligned} Q(\delta t) &= \sigma^2[1 - \exp\{-2\alpha_0(\delta t)\}] \\ &= \sigma^2[1 - \phi^2(\delta t)], \end{aligned} \tag{3.12}$$

which agrees with equation (3.5).

3.4 Some simulated error structures

The correlation function for an AR(1) error process, and three independent simulated sets of serially correlated errors taken at unequally spaced time points are shown in Fig. 3.1. The horizontal line in each simulation is the true mean of the observations. The unequally spaced time points were simulated as the sum of four independent exponentially distributed random variables, each with a mean of .025 time units, giving a mean separation of .1 time units. The time scales are not shown on the graphs since the graphs can be scaled to any dimensions. The parameters of the time series were chosen as

$$\begin{aligned} \alpha_0 &= 1.5 \\ v &= 0.25, \end{aligned}$$

so the variance of the error process is, from (3.11),

$$\sigma^2 = 0.25/(2 \times 1.5).$$

A sequence of standard Gaussian random variables was simulated. Denote the values of these random variables as z_j The first value was multiplied by σ so that it has the variance of the error process,

$$\epsilon_1 = \sigma z_1.$$

For a time step of δt, the autoregression coefficient was calculated as

$$\phi(\delta t) = e^{-1.5\delta t},$$

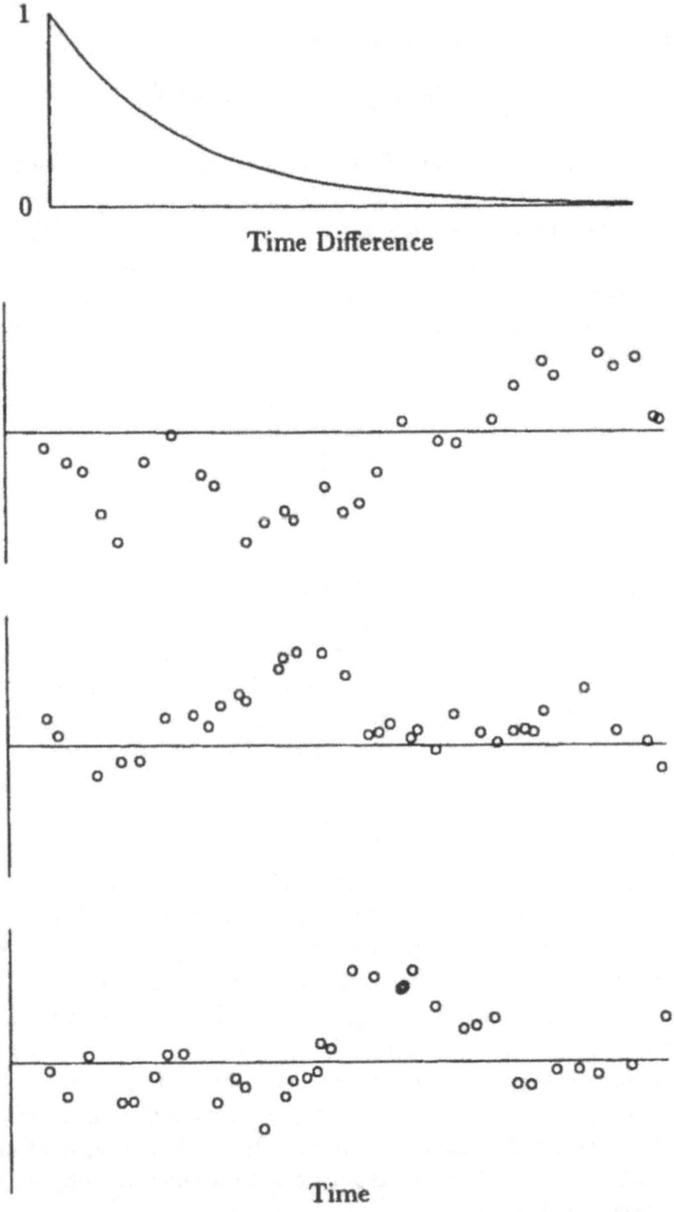

Fig. 3.1 *The correlation function of an AR(1) error structure and three independent simulated time series with AR(1) error structure at unequally spaced time points.*

and the value of the error process as

$$\epsilon_j = \phi(\delta t)\epsilon_{j-1} + z_j\sigma\sqrt{1-\phi^2(\delta t)}.$$

The standard deviation of the random input is obtained from (3.12).

Serial correlation is different from compound symmetry caused by the random subject effect (section 1.5). Compound symmetry occurs when the mean level varies from subject to subject and the within subject errors are uncorrelated. For this model, $\mathbf{Z}_i$ in (2.1) is a column of ones,

$$\mathbf{Z}_i = \begin{bmatrix} 1 \\ 1 \\ \vdots \\ 1 \end{bmatrix},$$

γ_i is a scalar, and $\mathbf{W}_i = \mathbf{I}$. Compound symmetry has constant correlation between all pairs of observations on a subject, and does not depend on the time separation between the observations as shown in the top part of Fig. 3.2. The random within subject error causes a discontinuity in the correlation function at zero. The correlation function jumps from 1 at time 0 down to the random subject effect variance divided by the sum of the random subject effect variance plus the random within subject error variance (1.6). Three simulations of random within subject errors and a random subject effect are shown in Fig. 3.2

When $\mathbf{W}_i$ has a serially correlated error structure and there is a random subject effect, the correlation decays with increasing time difference and approaches a limit that is greater than zero for large time differences. This limit is the intraclass correlation (1.6) produced by the random subject effect. The correlation function when there is both a random subject effect and within subject serial correlation is shown in the top part of Fig. 3.3. The correlation function is the top curve that approaches an asymptote at the level of the random subject effect. Three simulations of a serially correlated error structure with a random subject effect are shown in Fig. 3.3.

Adding random within subject error to a serially correlated structure provides more flexibility in fitting the error structure in a model. The addition of random within subject error causes this correlation structure to be discontinuous at zero as in the

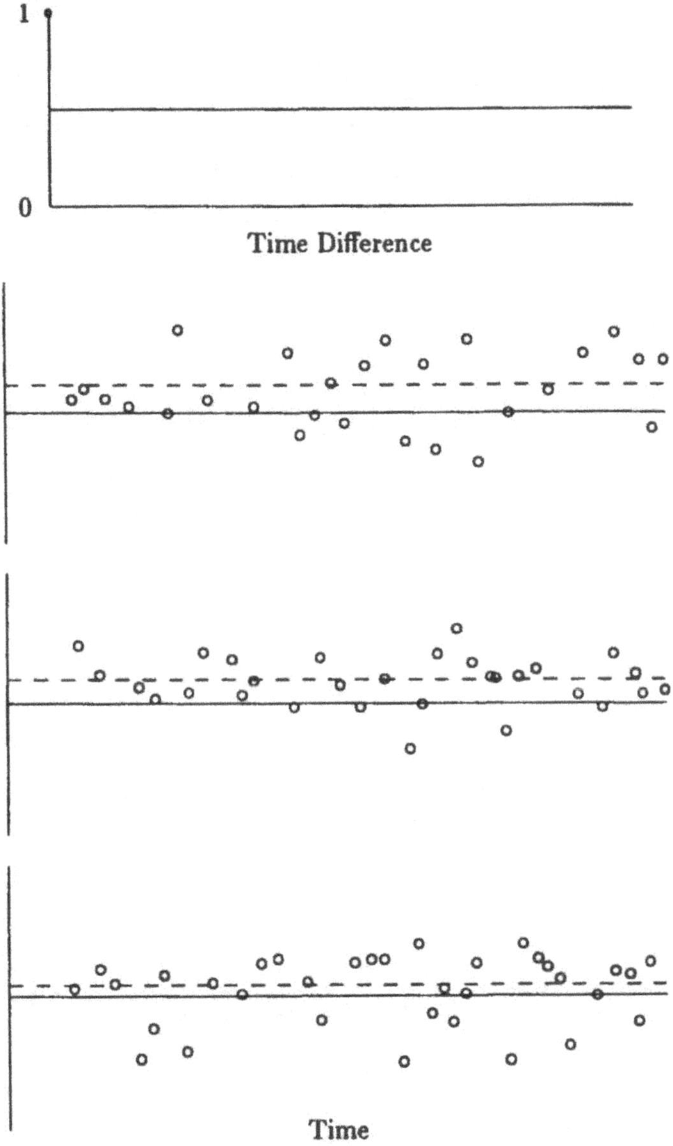

Fig. 3.2 *The correlation function of compound symmetry, a random subject effect with random within subject errors, and three independent simulated time series with this correlation structure at unequally spaced time points. The solid line is the group or population mean. The dashed line is the subject's mean.*

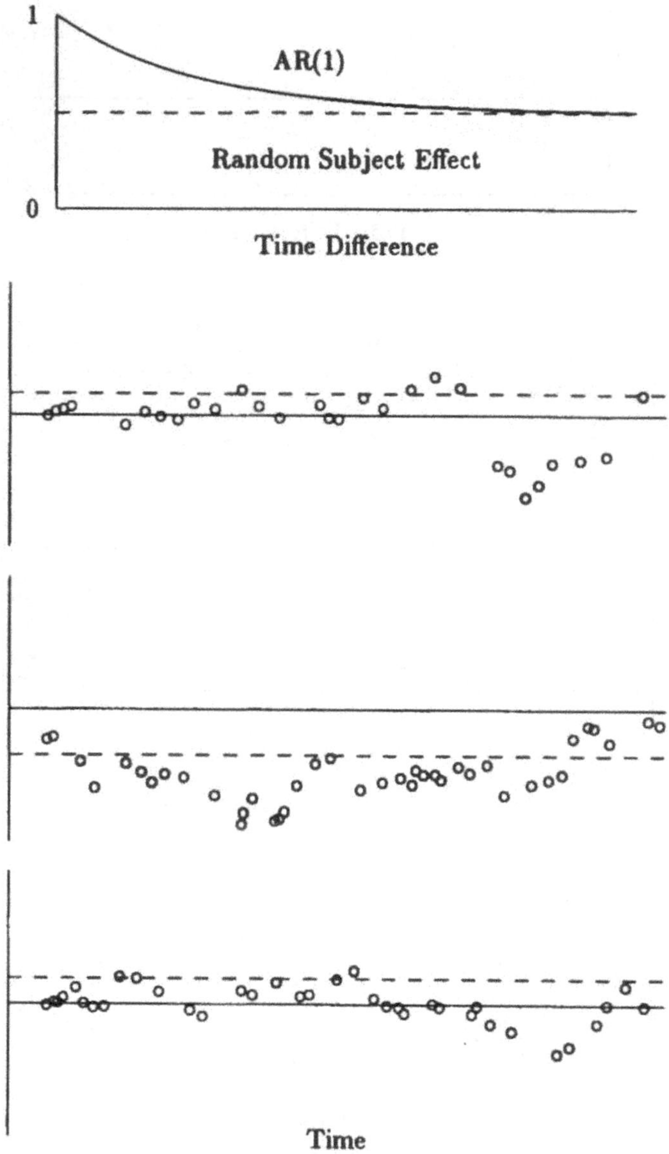

Fig. 3.3 *Correlation function of an AR(1) within subject error structure with a random effect, and three independent simulated time series with this covariance structure at unequally spaced time points. The solid line is the group or population mean. The dashed line is the subject's mean.*

top part of Fig. 3.4. Three simulations of a serially correlated error structures with a random subject effect and random within subject errors are shown in Fig. 3.4.

3.5 Other possible error structures

There are many possible structures for serial correlation, and the first order autoregressive structure, AR(1), is the simplest. Many data sets are not large enough to support complicated error structures. My philosophy is that the error structure is usually not simple, but with a moderate amount of data it is necessary to approximate reality. Introducing an AR(1) structure into the within subject errors may be far better than just assuming that the errors are uncorrelated, and it is possible to test statistically whether the AR(1) structure improves the fit.

A variance function can be introduced as in section 2.3. With serial correlation, the $\mathbf{W}_i$ matrix is not diagonal. The square root of the variance function associated with observation j on subject i can be used to multiply the j'th row and column of $\mathbf{W}_i$.

There are many situations where there appears to be serial correlation in the errors when random subject effects are not included in the model, and this apparent serial correlation goes away when random subject effects are included. A sensible approach to this problem is to fit four models, one with uncorrelated error structure, one with random subject effects alone, one with AR(1) within subject error structure alone, and one with both random subject effects and AR(1) error structure (Jones, 1990). This is a good situation for using AIC for model selection. There are four models which make six pairwise comparisons possible. The comparison of compound symmetry alone with AR(1) alone is a situation where the two models being compared are not nested so a likelihood ratio test cannot be used. They both have the same number of parameters and one model cannot be obtained from the other by constraining parameters. AIC can be used by calculating AIC for each of the four models (AIC is $-2 \ln$ likelihood plus twice the number of estimated parameters), and choosing the model for which AIC is lowest. It is possible for AIC to be negative, in which case the lowest AIC is the most negative AIC.

There are other approaches for modelling serial correlation in longitudinal data. Ante-dependence models introduced by Gabriel

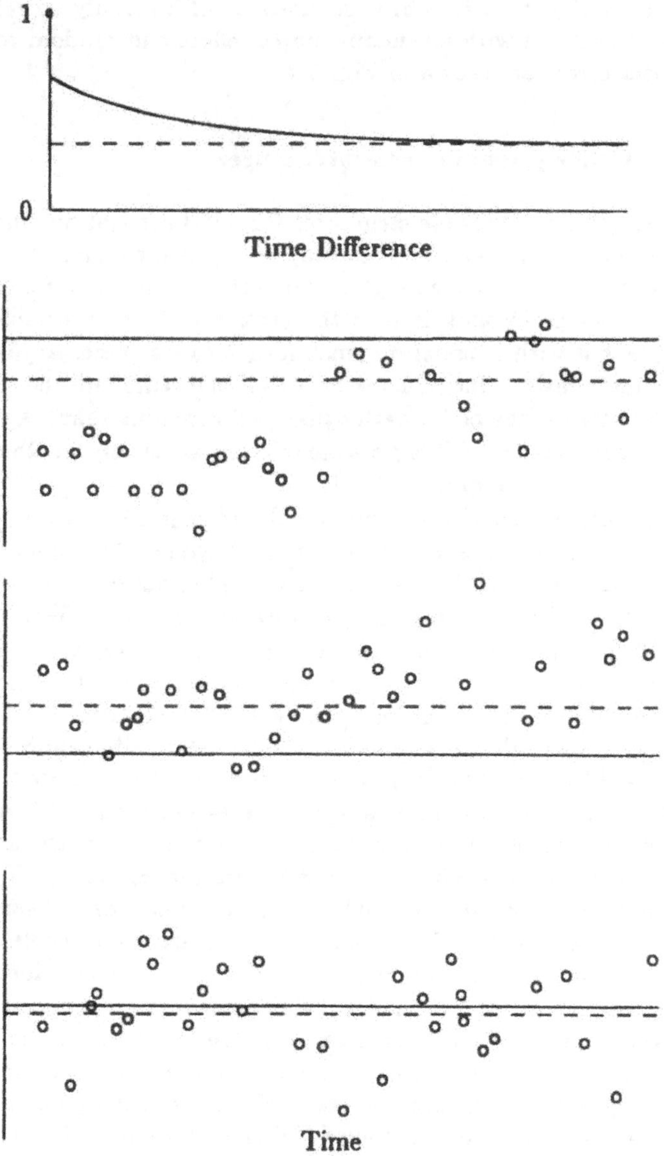

Fig. 3.4 *The correlation function of an AR(1) error structure with a random effect and random within subject errors, and three independent simulated time series with this correlation structure at unequally spaced time points. The solid line is the group or population mean. The dashed line is the subject's mean.*

(1962) and discussed by Kenward (1987) are related to autoregressive models, but are more general. Ante-dependence models are useful for balanced designs where the observations are not necessarily equally spaced in time. Consider the situation where each of the subjects is observed at the same n time points. An ante-dependence model of order one assumes that the within subject errors can have different variances at different times,

$$E(\epsilon_{ij}) = \sigma_j^2.$$

The within subject errors have an AR(1) type structure, but the autoregression coefficient can be different for each observation time,

$$\epsilon_{ij} = \phi_j \epsilon_{i,j-1}, \qquad \text{for} \qquad j > 1.$$

This model contains n variances and $n-1$ autoregression coefficients and generates a nonstationary error covariance structure. The covariance depends not only on the time difference between two observations, but also on where the observations occur in the sequence of observations. Ante-dependence models are useful for non-steady state situations such as subjects recovering from an injury or surgery. An AR(1) error structure is a special case where all the autoregression coefficients are determined by a single parameter, α_0, and all the variances are the same.

Higher order ante-dependence models are related to higher order autoregressions where the dependence goes back p observation times rather than one. In the limit, when the dependence goes back to the beginning of the sequence of observations, the ante-dependence model produces an arbitrary or unstructured within subject covariance matrix.

3.6 Circadian rhythms

Suppose observations are taken every four hours on a group of subjects, and it is of interest to know whether there is a daily fluctuation in the mean level with a period of 24 hours. There are several possible ways to formulate this problem. If the shape of the circadian rhythm is assumed to be completely general, the mean level can be modelled as taking different values at each observation time. Since there are six observations taken at the same times for each subject, the model would be

$$\begin{bmatrix} y_{i1} \\ y_{i2} \\ y_{i3} \\ y_{i4} \\ y_{i5} \\ y_{i6} \end{bmatrix} = \begin{bmatrix} 1 & 0 & 0 & 0 & 0 & 0 \\ 0 & 1 & 0 & 0 & 0 & 0 \\ 0 & 0 & 1 & 0 & 0 & 0 \\ 0 & 0 & 0 & 1 & 0 & 0 \\ 0 & 0 & 0 & 0 & 1 & 0 \\ 0 & 0 & 0 & 0 & 0 & 1 \end{bmatrix} \begin{bmatrix} \beta_1 \\ \beta_2 \\ \beta_3 \\ \beta_4 \\ \beta_5 \\ \beta_6 \end{bmatrix} + \begin{bmatrix} 1 \\ 1 \\ 1 \\ 1 \\ 1 \\ 1 \end{bmatrix} \gamma_i + \begin{bmatrix} \epsilon_{i1} \\ \epsilon_{i2} \\ \epsilon_{i3} \\ \epsilon_{i4} \\ \epsilon_{i5} \\ \epsilon_{i6} \end{bmatrix}. \tag{3.13}$$

The ϵ's may or may not be serially correlated. If the β's are estimated from a group of subjects and plotted on a time axis with a spacing of four hours, this would be the estimated circadian pattern for the group of subjects. To test the hypothesis that the pattern is significantly different from a constant,

$$H_0: \beta_1 = \beta_2 = \beta_3 = \beta_4 = \beta_5 = \beta_6,$$

the reduced model

$$\begin{bmatrix} y_{i1} \\ y_{i2} \\ y_{i3} \\ y_{i4} \\ y_{i5} \\ y_{i6} \end{bmatrix} = \begin{bmatrix} 1 \\ 1 \\ 1 \\ 1 \\ 1 \\ 1 \end{bmatrix} \beta_1 + \begin{bmatrix} 1 \\ 1 \\ 1 \\ 1 \\ 1 \\ 1 \end{bmatrix} \gamma_i + \begin{bmatrix} \epsilon_{i1} \\ \epsilon_{i2} \\ \epsilon_{i3} \\ \epsilon_{i4} \\ \epsilon_{i5} \\ \epsilon_{i6} \end{bmatrix}$$

can be fit to the data. The difference in the two values of -2 ln likelihood is tested as chi-square with 5 degrees of freedom. The null hypothesis is that there is no circadian rhythm. A large value of chi-square rejects the null hypothesis.

The hypothesis of no circadian rhythm can also be tested using a contrast matrix without fitting the reduced model. The hypothesis can be formulated

$$H_0: \begin{bmatrix} 1 & -1 & 0 & 0 & 0 & 0 \\ 0 & 1 & -1 & 0 & 0 & 0 \\ 0 & 0 & 1 & -1 & 0 & 0 \\ 0 & 0 & 0 & 1 & -1 & 0 \\ 0 & 0 & 0 & 0 & 1 & -1 \end{bmatrix} \begin{bmatrix} \beta_1 \\ \beta_2 \\ \beta_3 \\ \beta_4 \\ \beta_5 \\ \beta_6 \end{bmatrix} = \begin{bmatrix} 0 \\ 0 \\ 0 \\ 0 \\ 0 \end{bmatrix}$$

The above formulation assumes that the shape of the daily pattern is completely general. Also, if there are, say, hourly observations, there would be 24 coefficients to estimate. Usually there is some smoothness from hour to hour. The pattern does not jump up and down wildly. Fitting patterns consisting of sines and cosines can reduce the number of coefficients and give a variety of smooth patterns.

Assume that subject i is observed at times t_{ij}, $j = 1, n_i$. These observations can span a single day or many days. A general sine or cosine wave with a period of one day is (Zerbe and Jones, 1980)

$$A \cos\left(2\pi \frac{t_{ij} - \psi}{P}\right), \tag{3.14}$$

where A is the amplitude of the wave, P is the period, and ψ is the time of the maximum. The period of a circadian rhythm is one day, so P depends on the units of time. If time is measured in hours, $P = 24$. If time is measured in days, $P = 1$. It is usual to set the origin of time as midnight, at or before the first observation. If (3.14) function is fit to data, there are two parameters that need to be estimated, A and ψ. A is a linear parameter but ψ occurs nonlinearly in the function.

Using the formula for the cosine of the difference of two angles, this cosine wave can be written

$$\begin{aligned} A \cos\left(2\pi \frac{t_{ij} - \psi}{P}\right) &= A \cos\left(\frac{2\pi\psi}{P}\right) \cos\left(\frac{2\pi t_{ij}}{P}\right) \\ &+ A \sin\left(\frac{2\pi\psi}{P}\right) \sin\left(\frac{2\pi t_{ij}}{P}\right) \end{aligned}$$

This equation is of the form

$$\alpha \cos\left(\frac{2\pi t_{ij}}{P}\right) + \beta \sin\left(\frac{2\pi t_{ij}}{P}\right), \tag{3.15}$$

where

$$\begin{aligned} \alpha &= A \cos\left(\frac{2\pi\psi}{P}\right) \\ \beta &= A \sin\left(\frac{2\pi\psi}{P}\right). \end{aligned}$$

The inverse transforms are

$$\begin{aligned} A &= \sqrt{\alpha^2 + \beta^2} \\ \psi &= \frac{1}{2\pi} \arctan\left(\frac{\beta}{\alpha}\right). \end{aligned}$$

The fact that a general sine or cosine wave with a period of one day can be expressed as a cosine wave plus a sine wave with arbitrary coefficients is shown in Fig. 3.5. The top curve is a cosine

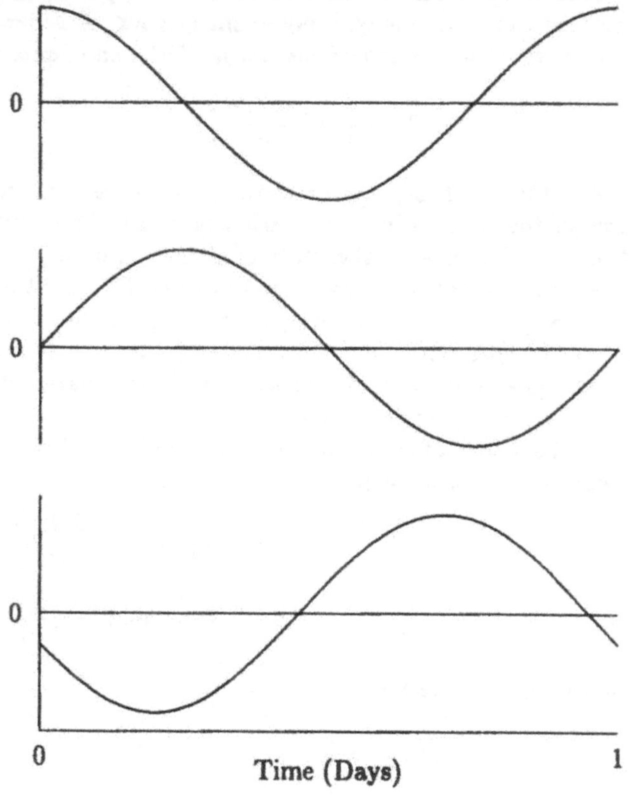

Fig. 3.5 *A cosine wave (top), a sine wave (middle) and the sum of the two with coefficients, -0.9631 and -0.3188*

wave with a period of one cycle per day, and the middle curve is a sine wave with a period of one cycle per day. The top curve was multiplied by -0.9631, and the middle curve was multiplied by -0.3188. The scaled curves were then added with the result being the bottom curve. To fit a sine wave to observations from a single subject, the coefficients α and β in (3.15) can be estimated by least squares producing a sine wave with arbitrary amplitude and phase. Arbitrary phase means that the maximum can occur anywhere within the period of the cycle.

To fit a sine wave to a group of subjects, the circadian rhythm model can be written

$$y_{ij} = \beta_1 + \beta_2 \cos\left(\frac{2\pi t_{ij}}{P}\right) + \beta_3 \sin\left(\frac{2\pi t_{ij}}{P}\right) + \gamma_i + \epsilon_{ij}. \quad (3.16)$$

β_1 is the mean level of the pattern, and β_2 and β_3 define the mean circadian pattern for the group. γ_i is a random subject effect which allows the mean level to vary randomly from subject to subject, and ϵ_{ij} are the within subject errors which can have an AR(1) structure. The fixed effects design matrix for subject i with t in hours

$$\mathbf{X}_i = \begin{bmatrix} 1 & \cos(2\pi t_{i1}/24) & \sin(2\pi t_{i1}/24) \\ 1 & \cos(2\pi t_{i2}/24) & \sin(2\pi t_{i2}/24) \\ & \vdots & \\ 1 & \cos(2\pi t_{in_i}/24) & \sin(2\pi t_{in_i}/24) \end{bmatrix}$$

The random effects design matrix, $\mathbf{Z}_i$ can be a column of ones which allows a random vertical shift from subject to subject. It is also possible for $\mathbf{Z}_i$ to be the same as the above $\mathbf{X}_i$ matrix. This would imply that each subject had a circadian pattern which randomly deviated from the mean pattern for the population. The disadvantage of this model is that the between subject covariance matrix $\mathbf{B}$ with upper triangular factor $\mathbf{U}$ has six elements which must be estimated nonlinearly. Usually this complication is not justified by the data, and a simple mean shift is sufficient.

The hypothesis of no circadian rhythm is

$$H_0: \ \beta_2 = 0 \quad \text{and} \quad \beta_3 = 0,$$

which can be tested by fitting the reduced or null model

$$y_{ij} = \beta_1 + \gamma_i + \epsilon_{ij} \quad (3.17)$$

and using a likelihood ratio test based on the change in -2 ln likelihood, which is tested as chi-square with two degrees for freedom. This hypothesis can also be tested using Wald's test based on the covariance matrix of the β's in the more general model. The contrast matrix is

$$\mathbf{C} = \begin{bmatrix} 0 & 1 & 0 \\ 0 & 0 & 1 \end{bmatrix}.$$

Wald's test also gives a chi-square test with two degrees of freedom, and for large sample sizes, the two test statistics should be close to each other.

A single sine wave with a period of one day may be too restrictive a model. Most circadian patterns do not follow a sine wave exactly. By adding higher harmonics to the model, more general patterns can be achieved. This is the Fourier series approach. The second harmonic has a period half as long as the fundamental frequency or two cycles per day. The third harmonic has three cycles per day, and so on. By adding more and more harmonics, an arbitrary cyclic pattern can be approached. However, with only a finite number of observations per day, harmonics can only be added until there are as many coefficients to be estimated as there are observations in a day. When this saturated model occurs, the Fourier series approach gives exactly the same results at the observation times as the general model used in (3.13).

The model with two harmonics is

$$\begin{aligned} y_i(t_{ij}) = \beta_1 &+ \beta_2 \cos\left(\frac{2\pi t_{ij}}{P}\right) + \beta_3 \sin\left(\frac{2\pi t_{ij}}{P}\right) \\ &+ \beta_4 \cos\left(\frac{4\pi t_{ij}}{P}\right) + \beta_5 \sin\left(\frac{4\pi t_{ij}}{P}\right) \\ &+ \gamma_i + \epsilon_{ij}. \end{aligned} \tag{3.18}$$

There must be at least five observations per day to use this model which is a much more general circadian pattern than (3.16). An example is shown in Fig. 3.6. A cosine wave and a sine wave of the fundamental frequency of one cycle per day are shown at the top. The fundamental frequency is also referred to as the first harmonic. The next two graphs are a cosine wave and a sine wave of the second harmonic with a frequency of two cycles per day, or a period of 12 hours. When these four curves are multiplied by constants and added together, results similar to the bottom curve are obtained. The bottom curve is a periodic function with a period of one day, but has a more general shape than a single cosine or sine wave. The principle of a Fourier series is that any periodic function can be approximated by a sum of cosine and sine waves consisting of a constant, the cosine and sine curves of the fundamental frequency multiplied by constants, the cosine and sine curves of the second harmonic multiplied by constants, plus higher harmonics (three cycles per day, four cycles per day,

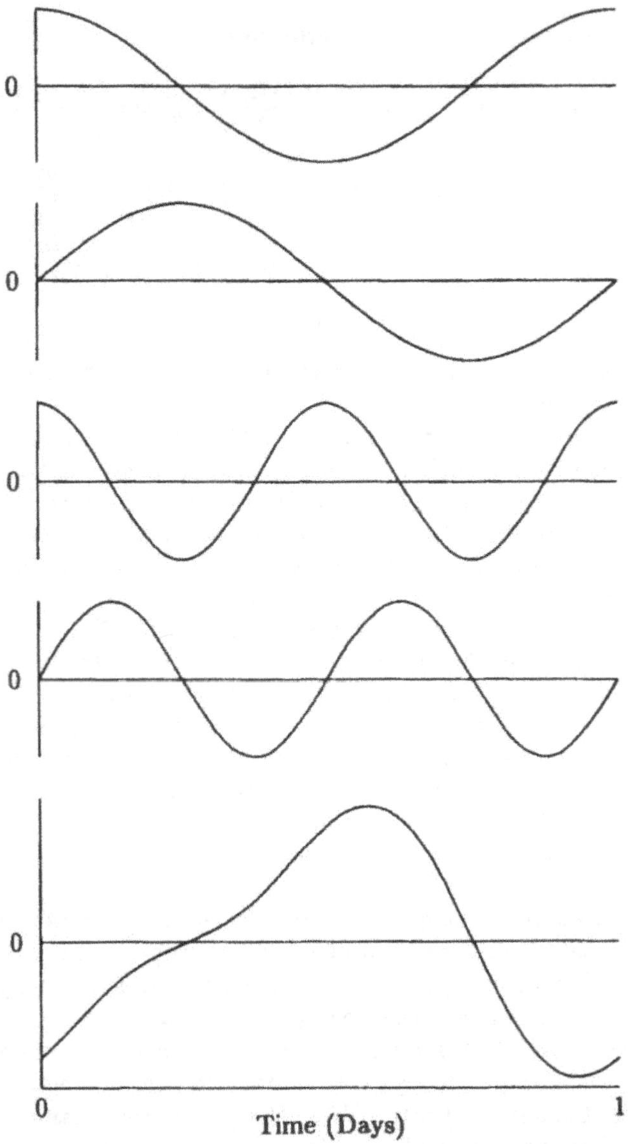

Fig. 3.6 *A cosine wave (top), and a sine wave (second) of the fundamental frequency of one cycle per day. A cosine wave (third) and a sine wave (fourth) of the second harmonic of two cycles per day. The sum of the four (bottom) with coefficients, -0.0045, -1.5115, 0.5071 and 0.0194.*

etc.). The model with three harmonics is

$$\begin{aligned}
y_i(t_{ij}) = \beta_1 &+ \beta_2 \cos\left(\frac{2\pi t_{ij}}{P}\right) + \beta_3 \sin\left(\frac{2\pi t_{ij}}{P}\right) \\
&+ \beta_4 \cos\left(\frac{4\pi t_{ij}}{P}\right) + \beta_5 \sin\left(\frac{4\pi t_{ij}}{P}\right) \\
&+ \beta_6 \cos\left(\frac{6\pi t_{ij}}{P}\right) + \beta_7 \sin\left(\frac{6\pi t_{ij}}{P}\right) \\
&+ \gamma_i + \epsilon_{ij}
\end{aligned}$$

To fit a periodic function to data, the constants can be estimated by least squares.

A natural way to approach the problem of choosing the number of harmonics to include in the model is to use a 'step up' approach. First, fit the null model (3.17), followed by the single sine wave model (3.16), then the model with two harmonics (3.18). At each step a likelihood ratio test can be calculated based on the change in $-2 \ln$ likelihood. The likelihood ratio test will have two degrees of freedom when comparing the null model (3.17) with the first harmonic model (3.16) since two β's have been added. Similarly, comparing (3.16) with (3.18) is a two degree of freedom test. AIC can also be calculated for each model. If there are eight or more observations per day, a model with three harmonics, which has seven coefficients, could be tried. Three harmonics are probably high enough for any number of observations per day. The more harmonics that are used, the rougher the circadian pattern can become.

A word of warning. While the Fourier series approach gives good results when the final fitted circadian pattern is calculated at the observation points, it may give poor results if the pattern is calculated between the observation points. This is a form of interpolation which should not be used. If higher harmonics are used, wild swings between the observation points sometime occur (Jones, Daniels and Bach, 1976). This phenomenon can also occur when fitting polynomials.

3.7 Exercises

1. A zero mean AR(1) process is

$$x_t = \phi x_{t-1} + \eta_t,$$

where

$$|\phi| < 1, \qquad E\{\eta_t\} = 0, \qquad \text{Var}\{\eta_t\} = \sigma_\eta^2,$$

η_s is uncorrelated with η_t for $s \neq t$, and η_t is uncorrelated with past values of x. Assuming ϕ is known, and observations are available up to time t, a one step prediction is

$$\hat{x}_{t+1} = \phi x_t,$$

and has variance σ_η^2. A two step prediction is

$$\hat{x}_{t+2} = \phi \hat{x}_{t+1}.$$

What is the prediction variance as you predict farther and farther into the future?

2. Prove that the covariance function of a stationary time series is a symmetric function of the lag (equation 3.3).

3. A zero mean AR(2) process is

$$x_t = \phi_1 x_{t-1} + \phi_2 x_{t-2} + \eta_t, \tag{3.19}$$

where

$$E\{\eta_t\} = 0, \qquad Var\{\eta_t\} = \sigma_\eta^2,$$

and η_s is uncorrelated with η_t for $s \neq t$, and η_t is uncorrelated with past values of x.

a) Calculate the Yule-Walker equation for $k = 0$ by multiplying equation (3.19) by x_t and taking expected values.

b) Calculate the Yule-Walker equations for $k > 0$ by multiplying equation (3.19) by x_{t-k} and taking expected values.

c) From the Yule-Walker equations for $k = 1$ and $k = 2$, what is the 2 by 2 system of equations that give the ϕ's in terms of the lag covariances?

d) From the Yule-Walker equation for $k = 0$, what is the equation that gives σ^2 in terms of the autoregression coefficients and the lag covariances?

e) Do these equations look like they have any relationship to the regression solution?

4. In the circadian rhythm model with two harmonics, (3.18), if there are six observations per day taken every four hours and a third harmonic is added to the model, only one new coefficient appears, and the model is equivalent to the general model (3.13). Why?

5. Simulate and plot a Wiener process by simulating a random walk with a very small time step and a very small variance for the random variable. Choose δt very small relative to the length of the x axis.

CHAPTER 4

State Space Representations

The direct approach for obtaining maximum likelihood (ML) or restricted maximum likelihood (REML) estimates of the $\mathbf{B}$ matrix and the parameters of the $\mathbf{W}_i$ matrix in the Laird-Ware model, (2.1), is to use initial guesses of the parameters to calculate the likelihood in equation (2.9) or the modified likelihood in equation (2.23). A nonlinear optimization program is used to search over the values of the unknown parameters to find the values that maximize the likelihood. The direct approach requires calculations with matrices of size $n_i \times n_i$, where n_i is the number of observations on subject i. These calculations are no problem when n_i is not large, and the use of the Cholesky factorization for calculating the likelihood is very stable numerically. This chapter considers an alternate method of calculating likelihoods using state space representations and the Kalman filter. State space methods use recursive calculations, entering the observations one at a time, and require calculations with small matrices the size of which do not depend on how many observations are available on a subject. Also, state space representations are very powerful in their own right. When a problem can be set up in state space form, the methodology exists for obtaining estimates of the unknown parameters. In Chapter 5 the Laird-Ware model for longitudinal data is given in state space form so that the Kalman Filter can be used to calculate exact likelihoods for Gaussian errors.

R. E. Kalman (1960) developed a recursive method of estimation and prediction of time series based on state space models. Within a year's time, this methodology was being used in the aerospace industry for estimating the position of rockets. It is a 'real time' estimation procedure allowing the estimates of the rocket's position to be continually updated as the data are collected. The term 'filter' is used when the estimation of the position and velocity of the rocket (the state of the system) is carried out at the time of the most recent set of observations. Since the observations have random errors, i.e. are noisy, improved estimates are obtained by using information about where the rocket was estimated to have been at the previous time and where it was predicted to be at the present time. This additional information uses what is known about the dynamics of the system. Kalman's methodology, as originally developed, should be classified under stochastic processes or probability since no parameters are being estimated. Variances and other parameters in the model are assumed known. This did not slow down applications of the models since engineers could usually produce reasonable values for the unknown parameters.

Schweppe (1965) showed that if the errors have a Gaussian distribution, the Kalman filter can be used to calculate likelihoods. Now we are in the realm of statistics since the ability to calculate likelihoods opens the door for maximum likelihood estimation of unknown parameters. The Kalman filter is not used in longitudinal data analysis as it is used in rocket tracking. It is used as a convenient way to calculate likelihoods for the purpose of obtaining maximum likelihood estimates of the parameters. The power of the Kalman filter approach to calculating likelihoods is that once a problem can be formulated in state space form, it is possible to calculate the likelihood recursively without using large matrices.

4.1 A tracking example

Consider a simple two dimensional 'rocket' example. A stone is thrown in the absence of air resistance. The only force acting on the stone is gravity. The 'state of the system' at a given time contains the information necessary to predict the future, assuming that the state is known exactly as are the forces acting on the system. In this example the state is the position and velocity of the stone. If the two dimensions are x and y, the state vector has four elements,

$$\mathbf{s}(t) = \begin{bmatrix} x(t) \\ \dot{x}(t) \\ y(t) \\ \dot{y}(t) \end{bmatrix},$$

where the dot over the letter indicates the derivative with respect to time. The state transition expresses how the state changes from one time point to the next. If the time interval between two observations is δt and there are no external forces acting on the system, the state transition equation would be

$$\begin{bmatrix} x(t) \\ \dot{x}(t) \\ y(t) \\ \dot{y}(t) \end{bmatrix} = \begin{bmatrix} 1 & \delta t & 0 & 0 \\ 0 & 1 & 0 & 0 \\ 0 & 0 & 1 & \delta t \\ 0 & 0 & 0 & 1 \end{bmatrix} \begin{bmatrix} x(t-\delta t) \\ \dot{x}(t-\delta t) \\ y(t-\delta t) \\ \dot{y}(t-\delta t) \end{bmatrix}.$$

The changes in the position coordinates equal the velocities multiplied by δt, the time interval. In the absence of outside forces, the velocities remain unchanged. Gravity, g, has dimensions distance per unit time squared, and if it is acting in the negative y direction, a constant vector can be added to the state equation,

$$\begin{bmatrix} x(t) \\ \dot{x}(t) \\ y(t) \\ \dot{y}(t) \end{bmatrix} = \begin{bmatrix} 1 & \delta t & 0 & 0 \\ 0 & 1 & 0 & 0 \\ 0 & 0 & 1 & \delta t \\ 0 & 0 & 0 & 1 \end{bmatrix} \begin{bmatrix} x(t-\delta t) \\ \dot{x}(t-\delta t) \\ y(t-\delta t) \\ \dot{y}(t-\delta t) \end{bmatrix} + \begin{bmatrix} 0 \\ 0 \\ -g\delta t^2/2 \\ -g\delta t \end{bmatrix}. \tag{4.1}$$

The last element of this vector is the change in velocity caused by the pull of gravity over a time interval of δt. The third element is the effect of this change integrated into the position (see Exercise 1). What has been developed here is a state equation with a fixed, deterministic, input vector based on a known force. Another vector can be added to the state equation for random inputs to the state equation such as may be caused by atmospheric turbulence. Equation (4.1) can be written in matrix notation as

$$\mathbf{s}(t) = \mathbf{\Phi}\mathbf{s}(t-\delta t) + \mathbf{f},$$

where

$$\mathbf{\Phi} = \begin{bmatrix} 1 & \delta t & 0 & 0 \\ 0 & 1 & 0 & 0 \\ 0 & 0 & 1 & \delta t \\ 0 & 0 & 0 & 1 \end{bmatrix},$$

and

$$\mathbf{f} = \begin{bmatrix} 0 \\ 0 \\ -g\delta t^2/2 \\ -g\delta t \end{bmatrix}.$$

A state space model has a second equation that tells what is actually observed. This is called the observation equation. Suppose that $x(t)$ and $y(t)$ are both observed at time intervals of δt but that there are observational errors. The observation equation expresses the observations as (hopefully) linear functions of the state vector. In this case the observation equation is

$$\begin{bmatrix} x_o(t) \\ y_o(t) \end{bmatrix} = \begin{bmatrix} 1 & 0 & 0 & 0 \\ 0 & 0 & 1 & 0 \end{bmatrix} \begin{bmatrix} x(t) \\ \dot{x}(t) \\ y(t) \\ \dot{y}(t) \end{bmatrix} + \begin{bmatrix} v_x(t) \\ v_y(t) \end{bmatrix}. \tag{4.2}$$

The observations (data) are denoted $x_o(t)$ and $y_o(t)$. The two positions are observed but the two velocities are not observed. It is not necessary to have observations on all the elements of the state vector. The observational errors are $v_x(t)$ and $v_y(t)$ and are assumed to have a Gaussian distribution with zero mean and covariance matrix $\mathbf{R}$. This 2×2 covariance matrix has 3 unique elements to be presumed known or estimated. If it is assumed that the two errors are independent with the same variance, the covariance matrix would be $\mathbf{R} = \sigma^2\mathbf{I}$ which only has one parameter to be presumed known or estimated. Perhaps σ^2 is a known property of the instrument used for the measurements. Equation (4.2) in matrix notation is

$$\mathbf{y}(t) = \mathbf{H}\,\mathbf{s}(t) + \mathbf{v}(t)$$

where

$$\mathbf{H} = \begin{bmatrix} 1 & 0 & 0 & 0 \\ 0 & 0 & 1 & 0 \end{bmatrix} \tag{4.3}$$

and

$$\mathbf{v}(t) = \begin{bmatrix} v_x(t) \\ v_y(t) \end{bmatrix}.$$

To start the recursive estimation procedure, it is necessary to know the initial values of the elements of the state vector at time zero, or have guesses with associated variances. When a rocket is on the launch pad, its position is known very precisely and its velocity is zero. If a stone is thrown, the initial position and velocity may not be known, but it is possible to guess at them and associate large variances with the guesses. Associating large variances with initial guesses indicates that the initial conditions are not really known and will be estimated from the data. Initial variances should not be extremely large since this may cause numerical problems, but reasonably large variances that indicate the lack of knowledge of the initial conditions should be used. This is the engineering approach: use the best possible guesses for the initial conditions, their standard deviations and the standard deviations of the errors. Experience has shown that if the guesses are approximately correct, reasonable estimates of the state vector as a function of time can be obtained along with the covariance matrices of the estimates.

The notation used for the estimates of the state vector is

$$\mathbf{s}(t|t-\delta t) = \begin{bmatrix} x(t|t-\delta t) \\ \dot{x}(t|t-\delta t) \\ y(t|t-\delta t) \\ \dot{y}(t|t-\delta t) \end{bmatrix},$$

which is the best estimate of the state at time t given observations up to time $t-\delta t$. This is a one step prediction based on the best estimate at time $t-\delta t$. Likewise,

$$\mathbf{s}(t|t) = \begin{bmatrix} x(t|t) \\ \dot{x}(t|t) \\ y(t|t) \\ \dot{y}(t|t) \end{bmatrix}$$

is the best estimate of the state at time t given observations up to time t. The covariance matrices of these two state vector estimates are

$$\mathbf{P}(t|t-\delta t) \quad \text{and} \quad \mathbf{P}(t|t).$$

The initial conditions are denoted $\mathbf{s}(0|0)$ with covariance matrix $\mathbf{P}(0|0)$.

The recursive procedure starts by assuming that the estimate of the state at time $t-\delta t$, using observations up to time $t-\delta t$, and the associated covariance matrix have been calculated,

$$\mathbf{s}(t-\delta t|t-\delta t) \quad \text{and} \quad \mathbf{P}(t-\delta t|t-\delta t).$$

It then proceeds through the calculation steps until $\mathbf{s}(t|t)$ and $\mathbf{P}(t|t)$ have been calculated. The starting point is from $\mathbf{s}(0|0)$ and $\mathbf{P}(0|0)$. From the estimate of where the stone was at time $t-\delta t$, a prediction of the state at time t is

$$\begin{bmatrix} x(t|t-\delta t) \\ \dot{x}(t|t-\delta t) \\ y(t|t-\delta t) \\ \dot{y}(t|t-\delta t) \end{bmatrix} = \begin{bmatrix} 1 & \delta t & 0 & 0 \\ 0 & 1 & 0 & 0 \\ 0 & 0 & 1 & \delta t \\ 0 & 0 & 0 & 1 \end{bmatrix} \begin{bmatrix} x(t-\delta t|t-\delta t) \\ \dot{x}(t-\delta t|t-\delta t) \\ y(t-\delta t|t-\delta t) \\ \dot{y}(t-\delta t|t-\delta t) \end{bmatrix} + \begin{bmatrix} 0 \\ 0 \\ -g\delta t^2/2 \\ -g\delta t \end{bmatrix}.$$

This equation in matrix notation is

$$\mathbf{s}(t|t-\delta t) = \mathbf{\Phi}\mathbf{s}(t-\delta t|t-\delta t) + \mathbf{f}.$$

Since this prediction is simply a linear combination of the state vector plus a constant vector, the covariance matrix is calculated as

$$\mathbf{P}(t|t-\delta t) = \mathbf{\Phi}\mathbf{P}(t-\delta t|t-\delta t)\mathbf{\Phi}'.$$

When there is random input in the state equation, this equation has another term on the end representing the covariance matrix of the random input.

The first and third elements of the predicted state vector are the predictions of the x and y coordinates of where the stone will be at time t. The $\mathbf{H}$ matrix, (4.3), expresses which elements of the state vector are being observed. The predicted position of the stone is

$$\mathbf{y}(t|t-\delta t) = \mathbf{H}\,\mathbf{s}(t|t-\delta t).$$

The difference between the predicted position of the stone and the observed position of the stone is called the *innovation*, in this case an innovation vector,

$$\mathbf{I}(t) = \mathbf{y}(t) - \mathbf{y}(t|t-\delta t).$$

The innovation is the important concept in the derivation of the Kalman filter. It is this concept that allows us to calculate likelihoods. The prediction of the current observations contains the information in the previous observations. The innovation vector contains the new information in the current observations that was unpredictable from the past, and the innovations are uncorrelated with, or orthogonal to, the past observations. If the observational errors are Gaussian, the innovations will be independent of the past, and the likelihood will be the product of the likelihoods of the innovations. To obtain the likelihood of an innovation vector, we need to know its covariance matrix. If there are no observational errors in the model, the covariance matrix of the innovation would be $\mathbf{HP}(t|t-\delta t)\mathbf{H}'$ since $\mathbf{y}(t|t-\delta t)$ is a linear transformation of $\mathbf{s}(t|t-\delta t)$ which has covariance matrix $\mathbf{P}(t|t-\delta t)$. Since the observational errors are independent of the past observations, the covariance matrix of the innovation vector is

$$\mathbf{V}(t) = \mathbf{HP}(t|t-\delta t)\mathbf{H}' + \mathbf{R}.$$

With Gaussian errors, the contribution to -2 ln likelihood from this innovation is

$$\ln|2\pi\mathbf{V}(t)| + \mathbf{I}'(t)\mathbf{V}^{-1}(t)\mathbf{I}(t).$$

This quantity is summed over all the observations to obtain the value of -2 ln likelihood for all the data.

To complete the Kalman filter recursion, it is necessary to update the estimate of the state vector and its covariance matrix based on the new information in the latest observations. The optimal estimate of the present value of the state vector based on the past observations is $\mathbf{s}(t|t-\delta t)$ with covariance matrix $\mathbf{P}(t|t-\delta t)$. The new information is in the innovation vector. The updated estimation is a weighted combination of the estimate based on the past and the new information,

$$\mathbf{s}(t|t) = \mathbf{s}(t|t-\delta t) + \mathbf{K}(t)\mathbf{I}(t),$$

where $\mathbf{K}(t)$, called the Kalman gain, is

$$\mathbf{K}(t) = \mathbf{P}(t|t-\delta t)\mathbf{H}'\mathbf{V}^{-1}(t).$$

The covariance matrix of this updated state vector is

$$\mathbf{P}(t|t) = \mathbf{P}(t|t-\delta t) - \mathbf{K}(t)\mathbf{HP}(t|t-\delta t).$$

As more observations are obtained, the velocity of the stone is estimated with increasing precision even though it is not directly observed. It is the state equation which incorporates the dynamics of the system that enables the unobserved components of the state vector to be estimated. Repeated observations of the position give information about the velocity.

If some of the parameters of the model are not known, maximum likelihood estimates of the parameters can often be obtained. Suppose that the observational error variance is $\mathbf{R} = \sigma^2\mathbf{I}$ and that σ^2 is unknown. By trying various values of σ^2 and calculating -2 ln likelihood using the Kalman recursion for each value, the value that gives the smallest -2 ln likelihood has the highest likelihood of the values tried. A systematic search algorithm would soon find the maximum likelihood estimate of σ^2. When there is only a single unknown parameter, it is not difficult to find the maximum likelihood estimate by simply trying different values until the minimum of -2 ln likelihood is found. With more than one unknown parameter, efficient nonlinear optimization programs exist for finding the values that minimize -2 ln likelihood (Dennis and Schnabel, 1983).

If an unknown parameter is a variance or an element of a covariance matrix, it is possible to concentrate it out of the likelihood so a search is not necessary over that parameter. If it is the only nonlinear parameter, it can be estimated with a single pass of the Kalman filter. In this case, all the variances in the model and the state covariance matrix are scaled by σ^2, so it is important not to attempt to concentrate a variance out of the likelihood which could have a maximum likelihood estimate of zero. An example of a variance whose maximum likelihood estimate could be zero is the between subject variance in models such as (1.1).

If the observational error covariance matrix is $\mathbf{R} = \sigma^2\mathbf{I}$, the scaling of all the variances and covariances implies that the state covariance matrix is $\sigma^2\mathbf{P}$ rather than $\mathbf{P}$. Now the innovation covariance is

$$\sigma^2\mathbf{V}(t) = \sigma^2\mathbf{H}\mathbf{P}(t|t-\delta t)\mathbf{H}' + \sigma^2\mathbf{R}.$$

The contribution to -2 ln likelihood is

$$\ln|2\pi\sigma^2\mathbf{V}(t)| + \frac{1}{\sigma^2}\mathbf{I}'(t)\mathbf{V}^{-1}(t)\mathbf{I}(t).$$

Summing this over all the observation times gives

$$\ell = \sum_{t=1}^{n}\left[\ln|2\pi\sigma^2\mathbf{V}(t)| + \frac{1}{\sigma^2}\mathbf{I}'(t)\mathbf{V}^{-1}(t)\mathbf{I}(t)\right].$$

If d observations are taken at every time point, then $\mathbf{V}(t)$ has dimensions $d \times d$. When a constant is moved from inside a determinant to outside, it must be raised to the power d since it multiplies every element of the matrix and the determinant consists of products of d elements. Now, if there are n observation times,

$$\ell = nd\ln(2\pi) + nd\ln(\sigma^2) + \sum_t \ln|\mathbf{V}(t)| + \frac{1}{\sigma^2}\sum_t \mathbf{I}'(t)\mathbf{V}^{-1}(t)\mathbf{I}(t). \tag{4.4}$$

In this equation, nd is the total number of observations, d observations at each of n time points. Taking the partial derivative of (4.4) with respect to σ^2 and setting the result equal to zero gives

$$\hat{\sigma}^2 = \frac{1}{nd}\sum_t \mathbf{I}'(t)\mathbf{V}^{-1}(t)\mathbf{I}(t),$$

and substituting this back into (4.4) gives −2 ln likelihood concentrated with respect to σ^2,

$$\ell = nd\ln(2\pi\hat{\sigma}^2) + \sum_t \ln|\mathbf{V}(t)| + nd.$$

To calculate this concentrated likelihood using the Kalman recursion,

$$\Delta = \sum_t \ln|\mathbf{V}(t)|$$

and

$$RSS = \sum_t \mathbf{I}'(t)\mathbf{V}^{-1}(t)\mathbf{I}(t)$$

are accumulated at each observation time. At the end of the data, σ^2 is estimated as

$$\hat{\sigma}^2 = \frac{1}{nd}RSS. \tag{4.5}$$

If σ^2 is the only unknown parameter in the model, this is the maximum likelihood estimate of σ^2. If there are other unknown parameters, −2 ln likelihood is

$$\ell = nd\ln(2\pi\hat{\sigma}^2) + \Delta + nd,$$

and this function is minimized with respect to the other unknown parameters to obtain their maximum likelihood estimates. This would be necessary if the stone throwing experiment were carried out on a different planet where the force of gravity, g, were unknown. The maximum likelihood estimate of σ^2 is calculated from (4.5) and is referred to as the mean square error (MSE) of the model.

4.2 The Kalman recursion

The usual state space model is more general than the example in the previous section. Random input is added to the state equation, and the various vectors and matrices may be time varying. This is important when the observations are unequally spaced since certain terms in the model will depend on the length of the time interval. The state equation is

$$\mathbf{s}(t) = \mathbf{\Phi}(t; t - \delta t)\mathbf{s}(t - \delta t) + \mathbf{f}_s(t) + \mathbf{G}(t)\mathbf{u}(t), \qquad (4.6)$$

and the observation equation is

$$\mathbf{y}(t) = \mathbf{H}(t)\,\mathbf{s}(t) + \mathbf{f}_o(t) + \mathbf{v}(t). \qquad (4.7)$$

In these equations $\mathbf{s}(t)$ is the state vector at time t, and $\mathbf{\Phi}(t; t-\delta t)$ is the state transition matrix from time $t - \delta t$ to time t. The vector $\mathbf{f}_s(t)$ represents non-random inputs to the state at time t. The random input to the state at time t is the vector $\mathbf{u}(t)$ which is assumed to have a Gaussian distribution which is independent of the past random inputs and of the observational errors. Because $\mathbf{u}(t)$ is premultiplied by the matrix $\mathbf{G}(t)$, without loss of generality, we can assume that $\mathbf{u}(t) \sim N(\mathbf{0}, \mathbf{I})$. Any correlation or scaling of the variances is handled by the matrix $\mathbf{G}(t)$. For example, the random input can have an arbitrary covariance matrix by allowing $\mathbf{G}(t)$ to be an arbitrary upper or lower triangular matrix. If the elements of $\mathbf{G}(t)$ do not vary with time and are estimated by nonlinear optimization, the covariance matrix of the random input is $\mathbf{GG}'$ which will always be a proper covariance matrix. It may be singular, but it cannot be negative. In the case of unequally spaced observations, it is $\mathbf{\Phi}(t; t-\delta t)$ and $\mathbf{G}(t)$ that depend on the spacing between observations.

In the observation equation, $\mathbf{y}(t)$ is the vector of observations at time t. The matrix $\mathbf{H}(t)$ shows which elements of the state vector are observed, or which linear combinations of the elements of the state vector are observed, and can be different at different times. An example is when there are several variables being observed at each time and there are missing values. In this case the vector of observations will change its length when there are missing observations, and the matrix $\mathbf{H}(t)$ will have the number of rows that corresponds to the number of actual observations. In this case, not only can the matrix $\mathbf{H}(t)$ change with time, its row dimension can change with time.

A vector of constant inputs to the observation equation, $\mathbf{f}_o(t)$, is also possible, and $\mathbf{v}(t)$ is the observational error vector which has a Gaussian distribution with a zero mean vector, independent at different times, independent of the random input to the state equation, and with covariance matrix $\mathbf{R}(t)$.

The recursion starts with the estimate of the state at time $t - \delta t$, $\mathbf{s}(t - \delta t|t - \delta t)$ and its covariance matrix $\mathbf{P}(t - \delta t|t - \delta t)$. The steps are:

1. Calculate a one step prediction,

$$\mathbf{s}(t|t - \delta t) = \mathbf{\Phi}(t; t - \delta t)\mathbf{s}(t - \delta t|t - \delta t) + \mathbf{f}_s.$$

2. Calculate the covariance matrix of this prediction,

$$\mathbf{P}(t|t-\delta t) = \mathbf{\Phi}(t; t-\delta t)\mathbf{P}(t-\delta t|t-\delta t)\mathbf{\Phi}'(t; t-\delta t) + \mathbf{G}(t)\mathbf{G}'(t).$$

In this equation, $\mathbf{G}(t)\mathbf{G}'(t)$ is the covariance matrix of the random input to the state equation over a time interval δt. This covariance matrix will often be denoted $\mathbf{Q}(\delta t)$, so the equation can also be written

$$\mathbf{P}(t|t - \delta t) = \mathbf{\Phi}(t; t - \delta t)\mathbf{P}(t - \delta t|t - \delta t)\mathbf{\Phi}'(t; t - \delta t) + \mathbf{Q}(\delta t).$$

3. Calculate the prediction of the next observation vector,

$$\mathbf{y}(t|t - \delta t) = \mathbf{H}(t)\mathbf{s}(t|t - \delta t) + \mathbf{f}_o.$$

4. Calculate the innovation vector which is the difference between the observation vector and the predicted observation vector,

$$\mathbf{I}(t) = \mathbf{y}(t) - \mathbf{y}(t|t - \delta t).$$

5. Calculate the covariance matrix of the innovation vector,

$$\mathbf{V}(t) = \mathbf{H}(t)\mathbf{P}(t|t - \delta t)\mathbf{H}'(t) + \mathbf{R}(t).$$

6. Accumulate the quantities needed to calculate −2 ln likelihood at the end of the recursion,

$$RSS \leftarrow RSS + \mathbf{I}'(t)\mathbf{V}^{-1}(t)\mathbf{I}(t),$$

$$\Delta \leftarrow \Delta + \ln|\mathbf{V}(t)|,$$

where $\leftarrow$ indicates that the left side is replaced by the right side as in a computer program.

7. To update the estimate of the state vector, let

$$\mathbf{A}(t) = \mathbf{H}(t)\mathbf{P}(t|t - \delta t),$$

then

$$\mathbf{s}(t|t) = \mathbf{s}(t|t - \delta t) + \mathbf{A}'(t)\mathbf{V}^{-1}(t)\mathbf{I}(t).$$

8. The updated covariance matrix is,

$$\mathbf{P}(t|t) = \mathbf{P}(t|t - \delta t) - \mathbf{A}'(t)\mathbf{V}^{-1}(t)\mathbf{A}(t).$$

Now return to step 1 until the end of the data is reached. If n is the total number of observations, i.e. all the elements of all the $\mathbf{y}(t)$ vectors (n replaces the nd in equation (4.4) since if there are missing observations, d may not be the same at every observation time), −2 ln likelihood is calculated as

$$\ell = n \ln(2\pi) + \Delta + RSS.$$

If one of the variances is concentrated out of the likelihood by setting its value in the recursion to 1 or any other constant, c, this variance is estimated as

$$\hat{\sigma}^2 = \frac{c}{n} RSS,$$

and −2 ln likelihood is

$$\ell = n \ln(2\pi\hat{\sigma}^2) + \Delta + n.$$

It must be emphasized that if there is a single observation at each observation time, n is the number of observations present. If there are one or more observations at each observation time, n is the total number of all these observations. It is possible to drop the 2π term from these equations since it appears in both forms of the likelihood. The n term on the end of the second form of −2 ln likelihood should not be dropped. If the same data set were used both with and without a variance concentrated out of the likelihood, the results would not be comparable if this n were dropped. I prefer to leave all constants in −2 ln likelihood so that published results in the literature can be compared. Since −2 ln likelihood can be negative, leaving the 2π term in the likelihood adds something positive so it is less likely to be negative. The value of −2 ln likelihood can be changed by multiplying the data by a constant (see the exercises).

The recursion can be started by specifying $\mathbf{s}(t_1|0)$ and $\mathbf{P}(t_1|0)$, where t_1 is the first observation time, and entering the recursion at the third step. If some of these initial conditions are unknown, they can be entered as unknown parameters to be estimated by nonlinear optimization.

4.2.1 *Missing observations*

If observations are equally spaced with some of the observations missing, there is a simple modification to the recursion. Typically a unique number which cannot appear in the actual data is stored at the times of the missing observations. After step 2 of the recursion, the observation at time t is tested to see if it is present or missing. If it is missing, set

$$\mathbf{s}(t|t) = \mathbf{s}(t|t - \delta t),$$

and

$$\mathbf{P}(t|t) = \mathbf{P}(t|t - \delta t),$$

and return to step 1. Since the calculations are usually carried out in place, i.e. $\mathbf{P}(t|t - \delta t)$ and $\mathbf{P}(t|t)$ occupy the same storage locations, there are no calculations involved at all. Simply return to step 1.

What is happening here is that a one step prediction of the state is calculated. If the observation at that time is missing, another one step prediction is made producing a two step prediction. The covariance matrix $\mathbf{P}(t|t - \delta t)$ will be larger for the two step prediction than for the one step prediction. This can be continued for any number of steps. If the process is stationary, where the effects of the remote past die out with time, the missing value procedure will approach a limit if there is a large gap of missing data. The effect will be the same as starting the recursion over at the end of the gap, although −2 ln likelihood will still be accumulated over all the available data. This philosophy also applies when there are independent realizations of the same process. The algorithm can be run over each segment with −2 ln likelihood accumulated over all the data. This is how multiple subjects are handled in longitudinal data.

4.3 AR(1) process with observational error

4.3.1 *Equally spaced data*

A zero mean AR(1) time series as shown in equation (3.1),

$$\epsilon_j = \phi\epsilon_{j-1} + \eta_j, \tag{4.8}$$

is already in the form of the state equation in a state space representation (4.6). The state vector is the scalar,

$$\mathbf{s}(j) = \epsilon_j,$$

the state transition matrix is now the scalar autoregression coefficient

$$\mathbf{\Phi}(j; j-1) = \phi,$$

and there are no non-random inputs,

$$\mathbf{f}_s(j) = 0.$$

There are several possibilities for the formulation of the random input to the state equation, η_j. It can be left as in (4.8) with η_j assumed to have an unknown variance, or η_j can be assumed to have unit variance and multiplied by G, which will then be the standard deviation of the random input. In this case the state equation is

$$\epsilon_j = \phi\epsilon_{j-1} + G\eta_j. \tag{4.9}$$

The initial conditions for the Kalman recursion are what is known about the state *in terms of the unknown parameters* (in this case, ϕ and G) before any data are collected. If the first observation is at $j = 1$, the initial state vector can be either $\epsilon(1|0)$ or $\epsilon(0|0)$. With equally spaced data it is usual to use $\epsilon(0|0)$. With unequally spaced data there is no natural time to be assigned 0, so $\epsilon(t_1|0)$ is often used. Since this is a zero mean AR(1) process, the initial state is

$$\epsilon(0|0) = 0.$$

The variance of this initial condition for the state is the variance of the process. From equation (3.5), the relation between the variance of the process, $C(0)$, and the variance of the random input is

$$\sigma^2 = C(0) = \frac{G^2}{1-\phi^2},$$

so the initial variance of the state is

$$P(0|0) = \frac{G^2}{1-\phi^2}.$$

Recall that it is necessary for ϕ to be in the range

$$-1 < \phi < 1$$

for the process to be stationary.

To concentrate a variance out of the likelihood, one of the variances in the model can be set to 1. There are two choices. The first is to set $P(0|0)$ equal to 1, in which case

$$G = \sqrt{1-\phi^2}.$$

In this case the mean square error (MSE) estimates $G^2/(1-\phi^2)$ so the estimate of G is

$$\hat{G} = \sqrt{(1-\hat{\phi}^2)MSE}$$

The second possibility is to set G to 1, in which case

$$P(0|0) = \frac{1}{1-\phi^2}.$$

In this case the mean square error estimates G^2.

So far, we have been considering only the state equation. For this to be a state space representation, we also need an observation equation. If the AR(1) process is observed without error, the observation equation is simply

$$y(j) = \epsilon_j. \tag{4.10}$$

The addition of random error to this equation makes the model more general. In the case of equally spaced data, this produces an autoregressive, moving average process of order 1,1 (ARMA(1,1)) (Box and Jenkins, 1976). With unequally spaced data, the addition of random error simply adds more flexibility to the model. It is a natural addition since random observational error often exists, sometimes in the form of round off error caused by rounding off the data to a given number of significant digits.

The observation equation with random error is

$$y(j) = \epsilon_j + v(j), \tag{4.11}$$

where $v(j)$ is the random error at time j assumed to have zero mean and variance R. If R is estimated by nonlinear optimization with a variance in the state equation concentrated out, the final estimate of R must be multiplied by MSE to obtain the actual estimate of the variance of the random error.

Unconstrained nonlinear optimization programs search for parameters over the whole real line. When parameters are restricted to an interval, transformations are necessary to keep the values of the parameters tried by the optimization program in bounds. If the observation error standard deviation is σ_o, the optimization can be carried out with respect to σ_o. In the subroutine written by the user to evaluate -2 ln likelihood, σ_o always appears squared,

$$R = \sigma_o^2.$$

Now whatever value is assigned to σ_o by the optimization program while searching for a minimum, R will always be non-negative. It is also possible to transform ϕ using a logistic type transformation as in Jones (1980),

$$\phi = \frac{1 - e^{-u}}{1 + e^{-u}}.$$

As u approaches $+\infty$, ϕ approaches 1, and as u approaches $-\infty$, ϕ approaches -1. When, $u = 0$ this transformation gives $\phi = 0$. The inverse transformation is

$$u = \ln \frac{1 + \phi}{1 - \phi}.$$

4.3.2 Arbitrarily spaced data

When the data are truly unequally spaced, i.e. not equally spaced with missing observations, various quantities in the recursion depend on the length of the time step. For a time step of length δt, the state equation is

$$\epsilon(t) = \phi(\delta t)\epsilon(t - \delta t) + \eta(\delta t), \qquad (4.12)$$

where $\eta(\delta t)$ has mean zero and variance

$$Q(\delta t) = \sigma^2[1 - \exp(-2\alpha_0 \delta t)]$$

from equation (3.12). The autoregressive coefficient, which depends on the time interval, is

$$\phi(\delta t) = \exp[-\alpha_0(\delta t)]$$

from equation (3.9). The parameter to be estimated is α_0. Since α_0 must be positive for stability, the natural transformation is a log transformation,

$$u_0 = \ln \alpha_0,$$

with the inverse transformation

$$\alpha_0 = \exp(u_0).$$

If observations are taken at times $t_1,\ t_2, \cdots, t_n$, the initial condition for the state could be

$$\epsilon(t_1|0) = 0.$$

The initial variance of the state is the variance of the process, σ^2,

$$P(t_1|0) = \sigma^2.$$

This is the natural variance to concentrate out of the likelihood, so if we set

$$P(t_1|0) = 1,$$

then

$$Q(\delta t) = [1 - \exp(-2\alpha_0 \delta t)],$$

and σ^2 is estimated by the mean square error (MSE). This would be consistent with the first choice in the equally space data case.

There is an interesting way to get started in the unequally space data case. In the above method, the recursion would be entered at step 3. Suppose we set the initial conditions to

$$\epsilon(0|0) = 0,$$

and

$$P(t_1|0) = 1,$$

and set the initial time step to zero ($\delta t = 0$). It is an easy exercise (see the Exercises) to show that if the recursion is entered at step 1 under these conditions, it is the same as entering the recursion at step 3 under the above conditions.

The observation equation for unequally spaced observations is

$$y(t_j) = \epsilon(t_j) + v(t_j),$$

where the $v(t_j)$ have constant variance R. This completes the setting up of an AR(1) process with observational error with unequally spaced data. It is the same as for equally space data except that ϕ and Q depend on the length of the time step.

A variance function (section 2.3), where the variance is proportional to the mean to some power, can be introduced here. This produces a nonstationary error structure. Using the mean from the previous iteration, both $Q(\delta t)$ and R are multiplied by this variance function to allow for variance heterogeneity. It is also necessary to multiply $P(t_1|0)$ by the value of this variance function at the first observation time.

4.4 Derivation of the Kalman recursion

All but the last two steps of this recursion are fairly intuitive. The derivation of these last two steps in Kalman's (1960) original paper is based on optimal estimates as conditional expectations. What follows is a simplified explanation of the derivation. The optimal estimate of the state at time t based on the data up to and including time t is the conditional expectation of the state given the observations,

$$\begin{aligned} \mathbf{s}(t|t) &= E\{\mathbf{s}(t)|\mathbf{y}(t), \mathbf{y}(t-\delta t), \cdots\} \\ &= E\{\mathbf{s}(t)|\mathbf{I}(t)\} + E\{\mathbf{s}(t)|\mathbf{y}(t-\delta t), \cdots\} \\ &= E\{\mathbf{s}(t)|\mathbf{I}(t)\} + \mathbf{s}(t|t-\delta t), \end{aligned}$$

since the innovation is uncorrelated with, or orthogonal to, the past observations. The optimal estimate of the state at time t can be written as a form of a regression equation,

$$\mathbf{s}(t) = \mathbf{K}(t)\mathbf{I}(t) + \mathbf{s}(t|t-\delta t) + \text{ error}, \tag{4.13}$$

where the conditional expectation has been replaced by unknown linear combinations of the innovation vector expressed as the matrix $\mathbf{K}(t)$, known as the Kalman gain.

In the geometrical approach to linear regression (Watson, 1967),

$$\mathbf{y} = \mathbf{X}\boldsymbol{\beta} + \epsilon$$

the normal equations can be derived by choosing $\boldsymbol{\beta}$ in such a way that the residuals are orthogonal to the columns of $\mathbf{X}$,

$$\mathbf{y} - \mathbf{X}\hat{\boldsymbol{\beta}} \perp \mathbf{X}.$$

Orthogonality in this setting is that the inner product (the sum of products of the elements) of two vectors must be zero, i.e.

$$\mathbf{X}'(\mathbf{y} - \mathbf{X}\hat{\boldsymbol{\beta}}) = \mathbf{X}'\mathbf{y} - \mathbf{X}'\mathbf{X}\hat{\boldsymbol{\beta}} = \mathbf{0},$$

which gives the normal equations to be solved for $\hat{\boldsymbol{\beta}}$,

$$\mathbf{X}'\mathbf{X}\hat{\boldsymbol{\beta}} = \mathbf{X}'\mathbf{y}.$$

In the Kalman filter setting, the inner product of two zero mean random variables is the expectation of their product, so orthogonality corresponds to zero correlation. In equation (4.13), $\mathbf{K}(t)$ is chosen such that

$$\mathbf{s}(t) - \mathbf{K}(t)\mathbf{I}(t) - \mathbf{s}(t|t - \delta t) \perp \mathbf{I}(t),$$

which is

$$E\{[\mathbf{s}(t) - \mathbf{K}(t)\mathbf{I}(t) - \mathbf{s}(t|t - \delta t)]\mathbf{I}'(t)\} = \mathbf{0}.$$

Since the covariation matrix of $\mathbf{I}(t)$ is $\mathbf{V}(t)$,

$$E\{[\mathbf{s}(t) - \mathbf{s}(t|t - \delta t)]\mathbf{I}'(t)\} - \mathbf{K}(t)\mathbf{V}(t) = \mathbf{0}.$$

Now, substituting from step 4 of the recursion for $\mathbf{I}(t)$, from the observation equation in the state space model for $\mathbf{y}(t)$, and from step 3 of the recursion for $\mathbf{y}(t|t - \delta t)$ gives

$$\mathbf{I}(t) = \mathbf{H}(t)\{\mathbf{s}(t) - \mathbf{s}(t|t - \delta t)\} + \mathbf{v}(t) \qquad (4.14)$$

$$E\{[\mathbf{s}(t) - \mathbf{s}(t|t - \delta t)][\mathbf{H}(t)\{\mathbf{s}(t) - \mathbf{s}(t|t - \delta t)\} + \mathbf{v}(t)]'\} = \mathbf{K}(t)\mathbf{V}(t).$$

Since $\mathbf{v}(t)$ is uncorrelated with the past errors and the errors in the state equation, and, since by definition,

$$\mathbf{P}(t|t - \delta t) = E\{[\mathbf{s}(t) - \mathbf{s}(t|t - \delta t)][\mathbf{s}(t) - \mathbf{s}(t|t - \delta t)]',$$

$$\mathbf{P}(t|t - \delta t)\mathbf{H}'(t) = \mathbf{K}(t)\mathbf{V}(t),$$

and the Kalman gain is

$$\mathbf{K}(t) = \mathbf{P}(t|t - \delta t)\mathbf{H}'(t)\mathbf{V}^{-1}(t).$$

The updated estimate of the state vector is

$$\mathbf{s}(t|t) = \mathbf{s}(t|t - \delta t) + \mathbf{K}(t)\mathbf{I}(t),$$

which is step 7 in the recursion.

The covariance matrix of the updated state is

$$\begin{aligned}\mathbf{P}(t|t) &= E\{[\mathbf{s}(t)-\mathbf{s}(t|t)][\mathbf{s}(t)-\mathbf{s}(t|t)]'\}\\ &= E\{[\mathbf{s}(t)-\mathbf{s}(t|t-\delta t)-\mathbf{K}(t)\mathbf{I}(t)]\\ &\quad\cdot[\mathbf{s}(t)-\mathbf{s}(t|t-\delta t)-\mathbf{K}(t)\mathbf{I}(t)]'\}\\ &= \mathbf{P}(t|t-\delta t)-\mathbf{K}(t)E\{\mathbf{I}(t)[\mathbf{s}(t)-\mathbf{s}(t|t-\delta t)]'\}-\\ &\quad E\{[\mathbf{s}(t)-\mathbf{s}(t|t-\delta t)]\mathbf{I}'(t)\}\mathbf{K}'(t)+\mathbf{K}(t)\mathbf{V}(t)\mathbf{K}'(t).\end{aligned}$$

Using (4.14),

$$E\{\mathbf{I}(t)[\mathbf{s}(t)-\mathbf{s}(t|t-\delta t)]'\}=$$
$$E\{[\mathbf{H}(t)\{\mathbf{s}(t)-\mathbf{s}(t|t-\delta t)\}+\mathbf{v}(t)][\mathbf{s}(t)-\mathbf{s}(t|t-\delta t)]'\}=$$
$$\mathbf{H}(t)\mathbf{P}(t|t-\delta t),$$

therefore,

$$\begin{aligned}\mathbf{P}(t|t) &= \mathbf{P}(t|t-\delta t)-\mathbf{K}(t)\mathbf{H}(t)\mathbf{P}(t|t-\delta t)\\ &\quad-\mathbf{P}(t|t-\delta t)\mathbf{H}'(t)\mathbf{K}'(t)+\mathbf{K}(t)\mathbf{V}(t)\mathbf{K}'(t).\end{aligned}$$

Substituting for $\mathbf{K}(t)$ shows that the last three terms are identical except for signs, so the updated state covariance matrix is

$$\mathbf{P}(t|t)=\mathbf{P}(t|t-\delta t)-\mathbf{K}(t)\mathbf{H}(t)\mathbf{P}(t|t-\delta t).$$

This is another expression for step 8 in the recursion.

4.5 Numerical considerations

The subtraction in step 8 of the recursion can cause the loss of significant digits in the state covariance matrix; therefore, the matrices $\mathbf{P}$, $\mathbf{V}$ and $\mathbf{A}$ should be double precision. Square root algorithms that work with a factored version of the $\mathbf{P}$ matrices, and are more stable numerically, do exist but use more computer time (Bierman, 1977). In step 7, the calculation of $\mathbf{A}(t)$ often involves a sparse $\mathbf{H}(t)$ matrix which may consist of ones and zeros. In this case the calculations can be made faster by not multiplying the two matrices, but by selecting the proper elements of $\mathbf{P}(t|t-\delta t)$ to put into $\mathbf{A}(t)$. Since covariance matrices are symmetric, only the upper triangular part of $\mathbf{V}(t)$, $\mathbf{P}(t|t-\delta t)$ and $\mathbf{P}(t|t)$ need be calculated, but care must be taken that the calculated elements be stored in the symmetric position before the matrix is used in a multiplication. Only calculating the upper triangular part of a covariance matrix saves a significant amount of time.

In the case of a vector of observations at each time, the recursion involves the inverse and determinant of $\mathbf{V}(t)$. These calculations can be easily carried out using the Cholesky factorization as in Chapter 3. The matrix $\mathbf{V}(t)$ is augmented by $\mathbf{A}(t)$ and by the innovation vector $\mathbf{I}(t)$ producing the matrix

$$[\ \mathbf{V}(t) \quad \mathbf{A}(t) \quad \mathbf{I}(t)\ ].$$

In the Cholesky factorization listed in Chapter 3, only the upper triangular part of $\mathbf{V}(t)$ need be stored, and the algorithm works in place replacing the original matrices by

$$[\ \mathbf{T} \quad \mathbf{D} \quad \mathbf{b}\ ],$$

where $\mathbf{T}$ is an upper triangular matrix such that

$$\mathbf{V}(t) = \mathbf{T}'\mathbf{T},$$

$$\mathbf{D} = (\mathbf{T}')^{-1}\mathbf{A}(t),$$

and

$$\mathbf{b} = (\mathbf{T}')^{-1}\mathbf{I}(t).$$

Now

$$\mathbf{I}'(t)\mathbf{V}^{-1}(t)\mathbf{I}(t) = \sum_k b_k^2,$$

the sum of squares of the elements of the vector $\mathbf{b}$, and

$$|\mathbf{V}(t)| = \prod_k T_{kk}^2,$$

the product of the squares of the diagonal elements of the triangular matrix $\mathbf{T}$. The factorization also simplifies the last two steps of the Kalman recursion. Step 7 can be calculated as

$$\mathbf{s}(t|t) = \mathbf{s}(t|t-\delta t) + \mathbf{D}'\mathbf{b},$$

and step 8 as

$$\mathbf{P}(t|t) = \mathbf{P}(t|t-\delta t) - \mathbf{D}'\mathbf{D}.$$

Factoring the matrix $\mathbf{V}(t)$ allows its inverse to be eliminated from the recursion and the actual inverse does not need to be calculated.

In addition to being fast, the Cholesky factorization is very stable numerically.

In the special case when the state vector has only one element, all the vectors and matrices in the recursion are scalars, the subtraction in step 8 of the recursion can be eliminated by algebra and this step becomes

$$P(t|t) = P(t|t - \delta t)R(t)/V(t). \tag{4.15}$$

Notice that if the observational error variance $\mathbf{R}(t)$ is zero, the variance of the estimate of the state at time t given data up to time t is zero. This is because a known constant $H(t)$ (usually 1 in this case) multiplied by the state is observed exactly, so the state is known exactly.

4.6 Exercises

1. Show that if the pull of gravity is in the negative y direction, the change in y from time $t - \delta t$ to time t is $\delta t\, \dot{y}(t - \delta t) - g\, \delta t^2/2$.

2. If the total number of observations is n and all the data are multiplied by 10, by how much does -2 ln likelihood change?

3. Set up a state space representation for a random walk observed at equally spaced time intervals with observational error, and write out the steps of the Kalman recursion for calculating -2 ln likelihood for given values of the unknown variances assuming independent Gaussian errors.

4. Set up a state space representation for a continuous time random walk observed at unequally spaced time intervals, and write out the steps of the Kalman recursion for calculating -2 ln likelihood for given values of the unknown variances assuming independent Gaussian errors. What changes would be necessary if one of the variances is concentrated out of the likelihood?

5. Verify equation (4.15) and find a simplified expression for step 7 in the Kalman recursion in the scalar case. Note the updated estimate of the state is a weighted average of the previous estimate and the new observation (with f_o subtracted).

6. Write out the steps of the Kalman recursion for the special case of an AR(1) process (4.9) with no observational error (4.10). Note that $P(j|j) = 0$ at all observation points.

7. Write out the steps of the Kalman recursion for the special case of an AR(1) process (4.9) with observational error (4.11).

8. Show that, with unequally spaced data, entering the recursion at step 1 with the initial $\delta t = 0$ is equivalent to entering the recursion at step 3.

CHAPTER 5

The Laird-Ware Model in State Space Form

In this chapter it will be shown how to set up the Laird-Ware model (2.1) in state space form. The Kalman filter can then be used to calculate the exact likelihood, and nonlinear optimization used to obtain maximum likelihood estimates. There are two concepts needed. The first is how to concentrate the fixed coefficients, $\boldsymbol{\beta}$, out of the likelihood so that only variance parameters need be estimated by nonlinear optimization. The second is how to handle the random effects.

The fixed effects are concentrated out of the likelihood by using the Kalman filter to generate the normal equations, for a weighted regression analysis, to estimate $\boldsymbol{\beta}$. To do this, the Kalman filter must be applied not only to the observation vectors, $\mathbf{y}_i$, but also to each column of the fixed effects design matrices, $\mathbf{X}_i$.

The random effects for each subject, $\boldsymbol{\gamma}_i$, are included in the state vector for the subject. The Kalman filter is reinitialized at the beginning of each subject's data, and the sums of squares and cross products necessary for the normal equations are summed across all the subjects. The maximum likelihood estimate of $\boldsymbol{\beta}$ is then calculated, *for the current values of the covariance parameters* that are being estimated by nonlinear optimization, and substituted back into the likelihood to obtain the likelihood concentrated with respect to $\boldsymbol{\beta}$.

5.1 Fixed effects

The use of the Kalman filter to calculate −2 ln likelihood when there are fixed effects in the model and autocorrelation in the errors would come under the heading *regression with stationary errors*. Harvey and Phillips (1979) used the Kalman filter for regression models with equally spaced autoregressive moving average (ARMA) errors by including the regression coefficients in the state vector. This treats the fixed effects as if they are random, but, by giving them large initial variances, the result is a Bayesian estimate of the fixed effects using a noninformative or diffuse prior (Box and Tiao, 1973, p. 32). A noninformative or diffuse prior is a Bayesian prior distribution which, in this case, has a uniform distribution from minus infinity to infinity. This type of prior is also called *improper* (Box and Tiao, 1973, p. 21) since it can not have area one under the probability distribution function. The reason for using noninformative priors is to base the statistical inference on the data rather than prior knowledge or subjective feelings. Using a noninformative prior produces restricted maximum likelihood (REML) estimates of the parameters of the model (Laird and Ware, 1982). Jones (1986) used a method that does not require increasing the length of the state vector by including the regression coefficients in the state vector. The fixed parameters are concentrated out of the likelihood. This method can be modified to give the REML estimates using equations (2.22) and (2.23).

For the general linear regression equation,

$$\mathbf{y} = \mathbf{X}\boldsymbol{\beta} + \boldsymbol{\epsilon}, \tag{5.1}$$

if the error vector has covariance matrix $\sigma^2\mathbf{V}$, the weighted least squares estimate is

$$\hat{\boldsymbol{\beta}} = (\mathbf{X}'\mathbf{V}^{-1}\mathbf{X})^{-1}\mathbf{X}'\mathbf{V}^{-1}\mathbf{y}. \tag{5.2}$$

A derivation of (5.2) is to multiply the regression equation by a lower triangular matrix which will produce transformed errors that are uncorrelated with constant variance. Let this lower triangular matrix be $\mathbf{L}$. Premultiplying (5.1) by $\mathbf{L}$,

$$\mathbf{Ly} = \mathbf{LX}\boldsymbol{\beta} + \mathbf{L}\boldsymbol{\epsilon}, \tag{5.3}$$

gives a new regression equation with a transformed vector of independent variables,

$$\tilde{\mathbf{y}} = \mathbf{Ly},$$

a transformed regression matrix,

$$\tilde{\mathbf{X}} = \mathbf{LX},$$

and a transformed error vector,

$$\tilde{\boldsymbol{\epsilon}} = \mathbf{L}\boldsymbol{\epsilon}.$$

The transformed regression equation is

$$\tilde{\mathbf{y}} = \tilde{\mathbf{X}}\boldsymbol{\beta} + \tilde{\boldsymbol{\epsilon}}. \tag{5.4}$$

Suppose the transformation is chosen such that the transformed errors have covariance matrix $\sigma^2\mathbf{I}$, i.e.

$$E\{\tilde{\boldsymbol{\epsilon}}\tilde{\boldsymbol{\epsilon}}'\} = \sigma^2\mathbf{I}.$$

But

$$E\{\tilde{\boldsymbol{\epsilon}}\tilde{\boldsymbol{\epsilon}}'\} = E\{\mathbf{L}\boldsymbol{\epsilon}\boldsymbol{\epsilon}'\mathbf{L}'\} = \mathbf{L}\,E\{\boldsymbol{\epsilon}\boldsymbol{\epsilon}'\}\mathbf{L}' = \sigma^2\mathbf{LVL}'.$$

Now $\mathbf{L}$ must be chosen such that

$$\mathbf{LVL}' = \mathbf{I},$$

i.e.

$$\mathbf{V} = \mathbf{L}^{-1}(\mathbf{L}')^{-1},$$

or

$$\mathbf{V}^{-1} = \mathbf{L}'\mathbf{L}.$$

This is a reverse Cholesky factorization of the inverse of the error covariance matrix in the sense that it is the product of an upper triangular matrix multiplied by its transpose rather than a lower triangular matrix multiplied by its transpose.

Consider now the Kalman filter operating on the error vector. Steps 1 and 3 of the Kalman recursion (page 87) that predict the state vector and the observation vector are linear operations on the state vector. Step 4, which calculates the innovation vector as the difference between the new observation vector and the predicted observation, is also a linear operation, as is step 7 which updates the estimate of the state vector. The innovation is the component of the new observation vector that is independent of the past and present observations, and is made up of a linear combination of the past and present observations. If the innovations are normalized by dividing by their standard deviations, $\sqrt{V(t_j)}$, the resulting sequence of normalized innovations can be represented by the lower triangular matrix $\mathbf{L}$ operating on the error vector. However, in regression, the error vector is not directly observable so the linear transformation is applied to the regression equation as in (5.3) giving the transformed regression equation (5.4). The implementation is that the Kalman filter is applied to $\mathbf{y}$ and to each column of the $\mathbf{X}$ matrix as if each of the columns is a data vector.

The solution of the transformed regression equation (5.4) is the usual least squares solution,

$$\hat{\boldsymbol{\beta}} = (\tilde{\mathbf{X}}'\tilde{\mathbf{X}})^{-1}\tilde{\mathbf{X}}'\tilde{\mathbf{y}}.$$

The total sum of squares (unadjusted for the mean) is

$$TSS = \tilde{\mathbf{y}}'\tilde{\mathbf{y}},$$

and the residual sum of squares is

$$\begin{aligned} RSS &= (\tilde{\mathbf{y}} - \tilde{\mathbf{X}}\hat{\boldsymbol{\beta}})'(\tilde{\mathbf{y}} - \tilde{\mathbf{X}}\hat{\boldsymbol{\beta}}) \\ &= \tilde{\mathbf{y}}'\tilde{\mathbf{y}} - \tilde{\mathbf{y}}'\tilde{\mathbf{X}}\hat{\boldsymbol{\beta}} = \tilde{\mathbf{y}}'\tilde{\mathbf{y}} - \tilde{\mathbf{y}}'\tilde{\mathbf{X}}(\tilde{\mathbf{X}}'\tilde{\mathbf{X}})^{-1}\tilde{\mathbf{X}}'\tilde{\mathbf{y}}. \end{aligned} \quad (5.5)$$

The estimate of σ^2 is

$$\hat{\sigma}^2 = \frac{1}{n} RSS$$

where n is the number of observations, and −2 ln likelihood is

$$\ell = n \ln(2\pi\hat{\sigma}^2) + \ln|\mathbf{V}| + n.$$

5.1.1 *The algorithm*

The Kalman recursion when fixed effects are concentrated out of the likelihood is similar to the recursion given on page 87 except that a state vector is needed for $\mathbf{y}$ and for each column of $\mathbf{X}$. If there are b columns of $\mathbf{X}$, the natural possibility is to have a state matrix with $b + 1$ columns. This state matrix will be denoted $\mathbf{S}(t)$, instead of the $\mathbf{s}(t)$ used for the state vector. Each column of $\mathbf{S}(0|0)$ is initialized the same as the single state vector would be initialized, and the state covariance matrix, $\mathbf{P}(0|0)$, is initialized as if there were no regression. It is assumed that a variance in the model is concentrated out of the likelihood by setting it equal to 1, and will be estimated by the mean square error. As is often the case in regression programs, a matrix should be set up in storage which is the $\mathbf{X}$ matrix of the regression augmented by the $\mathbf{y}$ vector,

$$[\ \mathbf{X} \quad \mathbf{y}\ ].$$

The rows of this matrix will be denoted

$$[\ \mathbf{x}(t_j) \quad \mathbf{y}(t_j)\ ].$$

It is necessary to initialize a scalar Δ and a $(b+1)\times(b+1)$ matrix $\mathbf{M}$ to zero for accumulating the determinant and sums of products for the regression. In the regression situation, there is a single observation corresponding to each row of the regression equation, so the innovation and the innovation covariance matrix, in that setting, become scalars. There is an innovation corresponding to each column of $\mathbf{X}$ and $\mathbf{y}$ so the innovation vector, rather than being a column vector as in the general Kalman recursion, will become a row vector with $b+1$ elements. The steps of the modified Kalman recursion which concentrates regression coefficients out of the likelihood are (see the appendix of Jones and Boadi-Boateng, 1991, for details in the case of unequally spaced AR(1) within subject errors):

1. Calculate a one step prediction,

$$\mathbf{S}(t_j|t_j-\delta t)=\mathbf{\Phi}(t_j;t_j-\delta t)\mathbf{S}(t_j-\delta t|t_j-\delta t).$$

   The only change in this step is that $\mathbf{\Phi}(t_j;t_j-\delta t)$ operates on a matrix rather than a vector, and there is no vector of constants, $\mathbf{f}_s$.

2. Calculate the covariance matrix of this prediction,

$$\begin{aligned}&\mathbf{P}(t_j|t_j-\delta t)\\&=\mathbf{\Phi}(t_j;t_j-\delta t)\mathbf{P}(t_j-\delta t|t_j-\delta t)\mathbf{\Phi}'(t_j;t_j-\delta t)+\mathbf{G}\mathbf{G}'.\end{aligned}$$

   This step is unchanged.

3. Calculate the prediction of the next observation vector,

$$\left[\ \mathbf{x}(t_j|t_j-\delta t)\quad y(t_j|t_j-\delta t)\ \right]=\mathbf{H}(t_j)\mathbf{S}(t_j|t_j-\delta t).$$

   A row vector of $b+1$ predictions is produced corresponding to the columns of $\mathbf{X}$ and $\mathbf{y}$, and there is no vector of constants, $\mathbf{f}_o$.

4. Calculate the innovation *row* vector which is the difference between the row of the $\mathbf{X}$ augmented by $\mathbf{y}$ matrix and the prediction of this row,

$$\mathbf{I}(t_j)=\left[\ \mathbf{x}(t_j)\quad \mathbf{y}(t_j)\ \right]-\left[\ \mathbf{x}(t_j|t_j-\delta t)\quad \mathbf{y}(t_j|t_j-\delta t)\ \right].$$

5. Calculate the variance of the innovation. This is the variance as if the Kalman filter is operating on the random error vector, ϵ,

$$V(t_j) = \mathbf{H}(t_j)\mathbf{P}(t_j|t_j - \delta t)\mathbf{H}'(t_j) + R(t_j).$$

This step is unchanged. The observational error variance, $R(t_j)$, is present only if there is observational error in the model.

6. Accumulate the quantities needed to calculate −2 ln likelihood at the end of the recursion,

$$\mathbf{M} \leftarrow \mathbf{M} + \mathbf{I}'(t_j)\mathbf{I}(t_j)/V(t_j),$$

$$\Delta \leftarrow \Delta + \ln V(t_j),$$

where ← indicates that the left side is replaced by the right side as in a computer program. To save time, only the upper triangular portion of the matrix **M** need be accumulated since it is a symmetric matrix and only the upper triangular portion will be used in the Cholesky factorization.

7. To update the estimate of the state vector, let

$$\mathbf{A}(t_j) = \mathbf{H}(t_j)\mathbf{P}(t_j|t_j - \delta t),$$

then

$$\mathbf{S}(t_j|t_j) = \mathbf{S}(t_j|t_j - \delta t) + \mathbf{A}'(t_j)\mathbf{I}(t_j)/V(t_j),$$

The fact that the innovation is now a row vector makes the state a matrix rather than a vector.

8. The updated covariance matrix is,

$$\mathbf{P}(t_j|t_j) = \mathbf{P}(t_j|t_j - \delta t) - \mathbf{A}'(t_j)\mathbf{A}(t_j)/V(t_j).$$

This step is unchanged except that $V(t_j)$ is now a scalar.

Now return to step 1 until the end of the data is reached.

The matrix **M** now contains

$$\begin{bmatrix} \tilde{\mathbf{X}}'\tilde{\mathbf{X}} & \tilde{\mathbf{X}}'\tilde{\mathbf{y}} \\ & \tilde{\mathbf{y}}'\tilde{\mathbf{y}} \end{bmatrix}.$$

The lower left block is left blank since it is not calculated. In fact, only the upper triangular part of the upper left block is calculated. If the full matrix were calculated, it would be symmetric and the lower triangular part would be filled in by simply storing the symmetric element in each lower triangular position. The upper portion of this matrix,

$$[\ \tilde{\mathbf{X}}'\tilde{\mathbf{X}} \quad \tilde{\mathbf{X}}'\tilde{\mathbf{y}}\ ],$$

is now subjected to a Cholesky factorization, page 186. This replaces the matrix by

$$[\ \mathbf{T} \quad \mathbf{b}\ ],$$

where $\mathbf{T}$ is an upper triangular matrix such that

$$\tilde{\mathbf{X}}'\tilde{\mathbf{X}} = \mathbf{T}'\mathbf{T},$$

and

$$\mathbf{b} = (\mathbf{T}')^{-1}\tilde{\mathbf{X}}'\tilde{\mathbf{y}}.$$

Now

$$\tilde{\mathbf{y}}'\tilde{\mathbf{X}}(\tilde{\mathbf{X}}'\tilde{\mathbf{X}})^{-1}\tilde{\mathbf{X}}'\tilde{\mathbf{y}} = \mathbf{b}'\mathbf{b},$$

the sum of squares of the elements of the vector $\mathbf{b}$. The residual sum of squares is now simply calculated as

$$RSS = \tilde{\mathbf{y}}'\tilde{\mathbf{y}} - \mathbf{b}'\mathbf{b}. \tag{5.6}$$

The first term on the right of the equal sign is available as the lower right hand element of the $\mathbf{M}$ matrix, and the second term is a simple sum of squares. The mean square error estimates the variance that is concentrated out of the likelihood,

$$MSE = \frac{1}{n} RSS$$

and −2 ln likelihood is

$$\ell = n \ln[2\pi(MSE)] + \Delta + n$$

as before.

A possible numerical difficulty in this algorithm is the subtraction in equation (5.6), which can cause the loss of significant digits, so these calculations should be carried out in double precision. There are other ways of avoiding this loss of precision that are more time consuming. The above procedure does not require the calculation of the regression coefficients. They can be calculated using the subroutine BACK, page 187 which solves a system of equations with a triangular coefficient matrix,

$$\mathbf{T}\hat{\boldsymbol{\beta}} = \mathbf{b}, \tag{5.7}$$

by back substitution. An exercise at the end of this chapter is to verify this equation. Now the weighted vector of residuals can be calculated,

$$\mathbf{e} = \tilde{\mathbf{y}} - \tilde{\mathbf{X}}\hat{\boldsymbol{\beta}}, \tag{5.8}$$

and the residual sum of squares is

$$RSS = \mathbf{e}'\mathbf{e}.$$

This method requires that the innovation vectors be saved or recalculated since, after dividing by the square root of the innovation variance, they contain the rows of $\tilde{\mathbf{X}}$ and the corresponding element of $\tilde{\mathbf{y}}$. It is faster to save the innovation vectors without normalization, and the innovation variances, calculate the residuals using (5.8), and calculate the weighted sum of squared of residuals,

$$RSS = \sum_{j=1}^{n} e(t_j)^2/V(t_j).$$

This avoids calculating the square roots of the variances. The two choices are to normalize the innovations or weight the squared residuals.

Another possibility suggested by Kitagawa (1987) uses Householder transformations to produce a QR factorization of the $\mathbf{X}_i$ matrices (Golub and Van Loan, 1983, p. 146). This method is more stable numerically than forming the normal equations and solving them using the Cholesky factorization, but requires more computing time.

5.1.2 Estimating an unknown mean

The easiest way to get a feeling for the state space formulation of the Laird-Ware model is to consider a simple example. Consider a model with an unknown mean and AR(1) errors,

$$y(t_j) = \beta + \epsilon(t_j) \tag{5.9}$$

where $\epsilon(t_j)$ is an AR(1) process, possibly with unequally spaced observations. There are several ways to formulate this problem in a state space representation. The state equation is (4.12)

$$\epsilon(t_j) = \phi(t_j - t_{j-1})\epsilon(t_{j-1}) + G(t_j - t_{j-1})\eta(t_j),$$

and the observation equation could be

$$y(t_j) = \epsilon(t_j) + \beta,$$

where β is the fixed function $\mathbf{f}_o(t_j)$ in the observation equation (4.7). The problem with this formulation is that β must be treated as an unknown parameter to be estimated by nonlinear optimization. The fewer parameters to be estimated by nonlinear optimization, the better.

Another possibility is to include the mean β in the state vector. The state equation would now be

$$\begin{bmatrix} \epsilon(t_j) \\ \beta \end{bmatrix} = \begin{bmatrix} \phi(t_j - t_{j-1}) & 0 \\ 0 & 1 \end{bmatrix} \begin{bmatrix} \epsilon(t_{j-1}) \\ \beta \end{bmatrix} + \begin{bmatrix} G(t_j - t_{j-1}) \\ 0 \end{bmatrix} \eta(t_j).$$

In this formulation there are problems starting off the recursion. If nothing is known about β or if an investigator does not want to introduce prior knowledge into the estimate, the initial value of the second element of the state could be set to zero with a large variance, essentially infinite. But how large is essentially infinite? If the initial variance is set too low, the final estimate will be biased towards zero since the final estimate will be a weighted average of the initial estimate and the estimated obtained from the data. A weighted average weights in proportion to the reciprocals of the variances. An initial estimate with a very large variance will have very little influence on the final result. However, a very large initial variance can cause numerical problems. It tends to 'swamp' the recursion and the subtraction in step 8 causes the loss of most of the significant digits. Ansley and Kohn (1985) have studied this problem in other applications and developed methods for starting Kalman filters with some of the initial variances infinite. They refer to this as em diffuse priors. Improved algorithms for diffuse priors have been developed by de Jong (1988, 1991a, 1991b).

The mean, β, is a fixed effect, and linear fixed effects can be concentrated out of the likelihood. This does not increase the length of the state vector and does not add another parameter to be estimated by nonlinear optimization. Writing equation (5.9) in vector form,

$$\mathbf{y} = \mathbf{1}\beta + \epsilon \tag{5.10}$$

puts it in the usual form for a regression equation, where $\mathbf{1}$ is a vector of ones which replaces the usual $\mathbf{X}$ matrix and β is the regression coefficient to be estimated. The usual unweighted estimate of β is the mean of the elements of the $\mathbf{y}$ vector. For a stationary error structure, such as AR(1) errors, the sample mean as an estimate of β is known to be asymptotically efficient. This means that if the number of elements of $\mathbf{y}$ is large, the sample mean will be very close to the maximum likelihood estimate. The question of what is large always comes up. If the autocorrelation between observations is near zero, 20 may be large enough; however, if the autocorrelation between observations is near one, 100 may not be large enough. This is because strong autocorrelation causes a deviation from the mean to persist over time so a small sample could catch all the observations on the same side of the true mean and thus give a poor estimate. The problem with using unweighted least squares (the sample mean) as an estimate is that the estimated standard error of the estimate is very biased. With positive autocorrelation, the estimated standard error tends to be low giving the impression that the mean is estimated more precisely than it actually is.

Since the error vector in this regression equation is not made up of uncorrelated elements but are autocorrelated, weighted least squares must be used to obtain the maximum likelihood estimate of β. If the error vector has covariance matrix $\sigma^2\mathbf{V}$, the weighted least squares estimate of β is

$$\hat{\beta} = (\mathbf{1}'\mathbf{V}^{-1}\mathbf{1})^{-1}\mathbf{1}'\mathbf{V}^{-1}\mathbf{y} = \frac{\mathbf{1}'\mathbf{V}^{-1}\mathbf{y}}{\mathbf{1}'\mathbf{V}^{-1}\mathbf{1}}. \tag{5.11}$$

If the observations are equally spaced without any missing observations, the model for the errors is as in equation (3.1) with the autoregression coefficient ϕ in the range $-1 < \phi < 1$. In this case the error covariance matrix is a Toeplitz matrix, constant along every diagonal,

$$\mathbf{V} = \begin{bmatrix} 1 & \phi & \phi^2 & \cdots & \phi^{n-2} & \phi^{n-1} \\ \phi & 1 & \phi & & \phi^{n-3} & \phi^{n-2} \\ \phi^2 & \phi & 1 & \ddots & \phi^{n-4} & \phi^{n-3} \\ \vdots & & \ddots & \ddots & \ddots & \vdots \\ \phi^{n-2} & \phi^{n-3} & \phi^{n-4} & \ddots & 1 & \phi \\ \phi^{n-1} & \phi^{n-2} & \phi^{n-3} & \cdots & \phi & 1 \end{bmatrix}. \tag{5.12}$$

The inverse of this matrix, $\mathbf{V}^{-1}$, is tridiagonal,

$$\frac{1}{1-\phi^2} \begin{bmatrix} 1 & -\phi & 0 & \cdots & 0 & 0 & 0 \\ -\phi & 1+\phi^2 & -\phi & & 0 & 0 & 0 \\ 0 & -\phi & 1+\phi^2 & \ddots & 0 & 0 & 0 \\ \vdots & & \ddots & \ddots & \ddots & & \vdots \\ 0 & 0 & 0 & \ddots & 1+\phi^2 & -\phi & 0 \\ 0 & 0 & 0 & & -\phi & 1+\phi^2 & -\phi \\ 0 & 0 & 0 & \cdots & 0 & -\phi & 1 \end{bmatrix}. \tag{5.13}$$

This may be verified by multiplication (see the Exercises). Note that this is not a Toeplitz matrix because of the ones in the upper left and lower right positions. This inverse matrix has a unique lower triangular factorization such that

$$\mathbf{V}^{-1} = \mathbf{L}'\mathbf{L},$$

$$\mathbf{L} = \frac{1}{\sqrt{1-\phi^2}} \begin{bmatrix} \sqrt{1-\phi^2} & 0 & 0 & \cdots & 0 & 0 & 0 \\ -\phi & 1 & 0 & & 0 & 0 & 0 \\ 0 & -\phi & 1 & & 0 & 0 & 0 \\ \vdots & & \ddots & \ddots & & & \vdots \\ 0 & 0 & 0 & \ddots & 1 & 0 & 0 \\ 0 & 0 & 0 & & -\phi & 1 & 0 \\ 0 & 0 & 0 & \cdots & 0 & -\phi & 1 \end{bmatrix}. \tag{5.14}$$

If equation (5.10) is premultiplied by $\mathbf{L}$, the least squares solution of this transformed regression equation is given by equation (5.11). If the transformed error vector,

$$\tilde{\epsilon} = \mathbf{L}\epsilon,$$

is Gaussian with zero mean and covariance matrix $\sigma^2\mathbf{I}$, this will be the maximum likelihood estimate.

The transformed errors are

$$\begin{aligned}
\tilde{\epsilon}_1 &= \epsilon_1 \\
\tilde{\epsilon}_2 &= (\epsilon_2 - \phi\epsilon_1)/\sqrt{1-\phi^2} \\
\tilde{\epsilon}_3 &= (\epsilon_3 - \phi\epsilon_2)/\sqrt{1-\phi^2} \\
\tilde{\epsilon}_4 &= (\epsilon_4 - \phi\epsilon_3)/\sqrt{1-\phi^2} \\
&\vdots \\
\tilde{\epsilon}_{n-1} &= (\epsilon_{n-1} - \phi\epsilon_{n-2})/\sqrt{1-\phi^2} \\
\tilde{\epsilon}_n &= (\epsilon_n - \phi\epsilon_{n-1})/\sqrt{1-\phi^2}.
\end{aligned}$$

These transformed errors all have variance σ^2, the variance of the errors, ϵ_j. This follows since the one step predictions,

$$\epsilon_j - \phi\epsilon_{j-1} = \eta_j,$$

have variance $(1-\phi^2)\sigma^2$ (3.5).

This transformation $\mathbf{L}$ is the Kalman filter. The Kalman filter is nothing other than the Cholesky decomposition. Pre multiplying by $\mathbf{L}$ generates the steps of the Kalman filter. This can be seen by working through the algorithm on page 87 with the following values:

$$\begin{aligned}
s(0|0) &= 0 \\
P(0|0) &= 1 \\
\Phi(j|j-1) &= \phi \\
f_s &= 0 \\
G &= \sqrt{1-\phi^2} \\
H(j) &= 1 \\
f_o &= 0 \\
R(j) &= 0
\end{aligned} \qquad (5.15)$$

Setting $R(j) = 0$ says there is no observational error, i.e. the AR(1) process is observed exactly at every time, so $P(j|j)$ will be zero at all observation times.

5.2 Random effects in the state

Duncan and Horn (1972) showed that mixed models can be put in state space form by including the random effects in the state vector. To put equation (2.1) in state space form, it is necessary to augment the state vector by the vector γ_i. The complete state space representation for subject i is

$$\begin{bmatrix} \epsilon_i(t_j) \\ \gamma_i(t_j) \end{bmatrix} = \begin{bmatrix} \phi(t_j - t_{j-1}) & 0 \\ 0 & 1 \end{bmatrix} \begin{bmatrix} \epsilon_i(t_{j-1}) \\ \gamma_i(t_{j-1}) \end{bmatrix} + \begin{bmatrix} G(t_j - t_{j-1}) \\ 0 \end{bmatrix} \eta_i(t_j). \qquad (5.16)$$

$$\xi_i(t_j) = \begin{bmatrix} 1 & \mathbf{z}(t_j) \end{bmatrix} \begin{bmatrix} \epsilon_i(t_j) \\ \gamma_i(t_j) \end{bmatrix} + v_i(t_j). \qquad (5.17)$$

Equation (5.16) indicates that the random effects for subject i are constant over time. The fact that the random effects vary from subject to subject is introduced into the model through the initial state covariance matrix which will be set equal to $\mathbf{B} = \mathbf{U}'\mathbf{U}$. Since the elements of this prior covariance matrix will be estimated by maximum likelihood, this is an empirical Bayes method. In equation (5.17), $\xi_i(t_j)$ represents the total error consisting of the within subject error, $\epsilon_i(t_j)$, and the between subject error $\mathbf{z}(t_j)\gamma_i$, where $\mathbf{z}(t_j)$ is a row vector consisting of row j in $\mathbf{Z}_i$ of equation (2.1).

Equations (5.16) and (5.17) are in the general form of a state space representation in equations (4.6) and (4.7), with

$$\mathbf{s}(t) = \begin{bmatrix} \epsilon_i(t) \\ \gamma_i \end{bmatrix},$$

$$\mathbf{\Phi}(t; t - \delta t) = \begin{bmatrix} \phi(\delta t) & 0 \\ 0 & 1 \end{bmatrix},$$

$$\mathbf{f}_s(t) = \mathbf{0}, \qquad (5.18)$$

$$\mathbf{G}(t) = \begin{bmatrix} G(\delta t) \\ 0 \end{bmatrix},$$

$$\mathbf{H}(t) = \begin{bmatrix} 1 & \mathbf{z}(t) \end{bmatrix},$$

and

$$\mathbf{f}_o(t) = \mathbf{0}.$$

When both fixed and random effects are included in the model, the modified Kalman recursion on page 104 can be used. With the variance of the within subject errors, $\epsilon(t)$, concentrated out of the likelihood, the initial state covariance matrix for each subject is

$$\mathbf{P}(t_1|0) = \begin{bmatrix} 1 & 0 \\ 0 & \mathbf{U}'\mathbf{U} \end{bmatrix}.$$

A variance function (section 2.3) can also be introduced here. The mean from the previous iteration can also include the subject specific random effects, so the estimated mean is the subject's mean. For observation j on subject i, let $\mathbf{x}_{ij}$ be the corresponding row of the $\mathbf{X}_i$ matrix and $\mathbf{z}_{ij}$ be the corresponding row of the $\mathbf{Z}_i$ matrix. The subject's mean level at that observation is $\mathbf{x}_{ij}\boldsymbol{\beta} + \mathbf{z}_{ij}\boldsymbol{\gamma}_i$. Estimates of $\boldsymbol{\beta}$ and $\boldsymbol{\gamma}_i$ are available from the previous iteration in the nonlinear optimization process. $G(\delta t)$, R and the upper left hand corner of $\mathbf{P}(t_1|0)$ are multiplied by this mean to some power to allow for variance heterogeneity.

5.2.1 *An example*

In a two group repeated measures experiments, such as (1.1), when there is no serial correlation within subjects, there exists an almost trivial state space representation. The state equation for subject i is

$$\gamma_i(t_j) = \gamma_i(t_{j-1}).$$

The random subject effect is constant within a subject since there is no random disturbance in this state equation. The initial state variance for this subject is

$$P(t_1|0) = U^2.$$

The observation equation, which is not really observed since there are fixed effects in the model, is

$$\xi_i(t_j) = \gamma_i(t_j) + \epsilon_{ij}.$$

In this formulation, since the ϵ's are independent, they have been moved into the observation equation to take the place of the random error. The variance of the ϵ's, σ^2, can be concentrated out of the likelihood by setting $R(t_j)$ in step 5 of the Kalman recursion (page 104) equal to 1. This leaves a single nonlinear parameter, U. In the balanced case when there are the same number of observations on every subject, U can be estimated in closed form as was shown in Chapter 2. In the unbalanced case, the maximum likelihood estimate of U can be obtained by searching over values of U for the minimum of -2 ln likelihood. The β's can be concentrated out of the likelihood by noting that one possible set of $\mathbf{X}_i$ matrices for the regression is

$$\mathbf{X}_i = \begin{bmatrix} 1 & 0 \\ 1 & 0 \\ \vdots & \\ 1 & 0 \end{bmatrix}$$

for subjects in the first group, and

$$\mathbf{X}_i = \begin{bmatrix} 0 & 1 \\ 0 & 1 \\ \vdots & \\ 0 & 1 \end{bmatrix}$$

for subjects in the second group. The two estimated regression coefficients are the group means, β_k.

A second formulation of this example is to keep the within subject errors in the state equation,

$$\begin{bmatrix} \epsilon_i(t_j) \\ \gamma_i(t_j) \end{bmatrix} = \begin{bmatrix} 0 & 0 \\ 0 & 1 \end{bmatrix} \begin{bmatrix} \epsilon_i(t_{j-1}) \\ \gamma_i(t_{j-1}) \end{bmatrix} + \begin{bmatrix} 1 \\ 0 \end{bmatrix} \eta_i(t_j). \tag{5.19}$$

$$\xi_i(t_j) = \begin{bmatrix} 1 & 1 \end{bmatrix} \begin{bmatrix} \epsilon_i(t_j) \\ \gamma_i(t_j) \end{bmatrix}. \tag{5.20}$$

This is the limiting case of the AR(1) within subject error structure as the autoregression coefficient approaches zero. However, for unequally spaced data using a continuous time AR(1) structure, this form can only be approached, not reached, since in equation (5.16),

$$\phi(t_j - t_{j-1}) = \exp[-\alpha_0(t_j - t_{j-1})],$$

α_0 must go to infinity for $\phi(t_j - t_{j-1})$ to go to zero. To test the hypothesis of independent within subject errors against the alternative of continuous time AR(1) errors, model (5.19) must be fit as a special case. The advantage of this formulation is that a special program does not need to be written for this case. It is important to note that there is no random error in the observation equation (5.20). If observational error were included in the model, it would be confounded with the independent error in the state equation.

5.3 Predictions

Predicting the population mean curves and individual predictions for subjects for which there have been no data collected are the same as in section 2.7, page 42. For individual predictions for a subject who has previous observations, serial correlation in the within subject errors can be used to improve the predictions. A byproduct of the state space method of estimation is an estimate of the random effects for the individual subjects together with the covariance matrix of these random effects. These are the same empirical Bayes estimates obtained by Laird and Ware (1982) using the EM algorithm, except that the covariance matrix has not been adjusted for the estimation of $\boldsymbol{\beta}$. The estimates produced by the Kalman filter are the same as in equation (2.24),

$$\hat{\gamma}_i = \mathbf{B}\mathbf{Z}_i'\mathbf{V}_i^{-1}(\mathbf{y}_i - \mathbf{X}_i\hat{\boldsymbol{\beta}}),$$

with covariance matrix

$$\mathrm{cov}(\hat{\gamma}_i - \gamma_i) = \hat{\sigma}^2(\mathbf{B} - \mathbf{B}\mathbf{Z}_i'\mathbf{V}_i^{-1}\mathbf{Z}_i\mathbf{B}), \qquad (5.21)$$

which is (2.35) without the last term. The last term is the addition to the covariance matrix caused by estimating the fixed regression coefficients, $\boldsymbol{\beta}$.

After maximum likelihood estimates of the parameters are obtained, the maximum likelihood estimates of $\boldsymbol{\beta}$, (5.7), can be used to subtract the fixed effects regression from the observations for each subject. The residuals for subject i are

$$\mathbf{e}_i = \mathbf{y}_i - \mathbf{X}_i\hat{\boldsymbol{\beta}}.$$

Using (5.16) as the state equation, these residuals can now be used as the observations, $y(t)$, in the Kalman recursion (page 87). Note that the state vector is of length $g + 1$, where g is the length of the vector of random effects, γ_i. The first element corresponds to the AR(1) error structure, and the last g elements to the random effects. At the end of each subject's data, the last g elements of the state vector contain the estimate of that subject's random effects, and the lower right $g \times g$ block of the state covariance matrix, $\mathbf{P}(t_{n_i}|t_{n_i})$, contains the covariance matrix of the subject's random effects.

The special case of the Kalman filter for these predictions, (5.18), starts with the elements of the state set to zero and

$$\mathbf{P}(t_1|0) = \hat{\sigma}^2 \begin{bmatrix} 1 & 0 \\ 0 & \mathbf{U}'\mathbf{U} \end{bmatrix}.$$

Variances and covariances are scaled back to their original units by including $\hat{\sigma}^2$ in the covariance matrices. Step 2 of the Kalman filter (page 87) becomes

$$\mathbf{P}(t|t-\delta t) = \mathbf{\Phi}(t;t-\delta t)\mathbf{P}(t-\delta t|t-\delta t)\mathbf{\Phi}'(t;t-\delta t)+\hat{\sigma}^2\mathbf{G}(t)\mathbf{G}'(t).$$

Again, the random input is scaled by $\hat{\sigma}^2$. Step 5 of the Kalman filter is also modified,

$$\mathbf{V}(t) = \mathbf{H}(t)\mathbf{P}(t|t-\delta t)\mathbf{H}'(t) + \hat{\sigma}^2\mathbf{R}(t) + \mathbf{x}'\text{cov}(\hat{\boldsymbol{\beta}})\mathbf{x}.$$

The variance of the random input is scaled by $\hat{\sigma}^2$ and a term is added to allow for the increase in variance caused by the estimation of $\boldsymbol{\beta}$. Step 6 of the Kalman filter is omitted since the likelihood is not being calculated.

5.4 Shrinkage estimates

Before data are observed on a subject, the random effects have the value zero and the prior covariance matrix $\sigma^2\mathbf{B}$. After the observations of the subject, the random effects have the posterior covariance matrix consisting of the lower right hand block of $\mathbf{P}(t_{n_i}|t_{n_i})$. The estimates of the random effects consist of weighted averages of the prior values of zero and the values that would have been obtained from the data alone. This weighted average causes the estimates of the random effects to be shrunk towards zero from the values that would have been obtained from the observations alone. This is often referred to as Stein shrinkage (Stein, 1956). For example, in simple repeated measurements of blood pressure on subjects, one method of estimating the random effect for each subject is to average the subject's observation and subtract the mean of all the subjects in the group. In a balanced design where each subject has the same number of observations and there is no serial correlation, the mean of all the observations on all the subjects in the group is the estimate of the fixed effects regression coefficient. The estimates obtained from the output of the Kalman filter will be closer to zero than the estimates obtained from simply averaging the observations for each subject. The amount of this shrinkage depends on the relative values of the within subject and between subject variances. In the extreme case of the between subject variance σ_γ^2 equal to zero, there is no random effect in the model and the estimates of the random effects for each subject are zero. This is complete shrinkage, all the way to zero. Every subject's mean is estimated as the grand mean of all the subjects. In the other extreme where the between subject variance is very large compared to the within subject variance, σ_ϵ^2, the estimates of the random effects will be very close to the subjects average minus the group average.

The basic concept here is the estimates for each subject borrows information from the other subjects in the group. In the extreme case of no random effects, every subject has the same mean level so the best estimate of this is the group mean. At the other extreme, the means vary wildly from subject to subject so the best estimate of a subject's level is that subject's mean. Most cases fall between these extremes and the best estimate of a subject's level is affected by the group level.

5.5 Data analysis

The error structure of many data sets, especially if they are not too large, can be modelled using random effects and within subject AR(1) error structure with observational error. Small data sets cannot usually support more complicated within subject error structures, and assuming an AR(1) structure may be much better than assuming independent within subject errors. The assumption of an AR(1) error structure versus the alternative of independent within subject errors can be tested by fitting both models and using the likelihood ratio test based on the change in $-2 \ln$ likelihood. For large data sets, it may be worth while trying more complicated within subject error structures than AR(1) with observational error. These models are considered in the next chapter.

An approach to data analysis is to first determine the most complete model for the fixed effects to be considered (Jones, 1990). In a balanced design, often this can be the saturated model where every time point for every group has a different level. Effort can then be directed towards determining a good model for the random between subject effects. A single random effect for each subject indicates a random shift in level from subject to subject. If the fixed effects are straight lines or higher order polynomials, the random effects could have two components for each subject indicating a random deviation straight line for each subject. This means that each subject's data, apart from purely random error, fall on a straight line with a deviation from the group curve that has a random intercept and slope with zero means and arbitrary covariance matrix. Again, two random effects versus one can be tested by fitting both models and using the likelihood ratio test. After a good model is determined using the fixed and random effects, AR(1) error structure can be included to see if there is significant improvement in the fit. After the error model has been determined, the fixed effects can be varied to test the main hypotheses of interest.

An example of a large unbalanced data set, where an AR(1) error structure with observational error significantly improves the fit over random subject effects alone, is given in Gabow et al. (1992). In this application, the random subject effect consisted of a straight line for each subject with a random slope and intercept.

5.6 Exercises

1. Verify that the solution of the triangular system of equations in equation (5.7) is the proper weighted least squares estimate of the regression coefficients.

2. Verify by multiplication that equation (5.13) is the inverse of equation (5.12).

3. Verify that (5.14) is the lower triangular factor of $\mathbf{V}^{-1}$.

4. Verify that the least squares solution of equation (5.10) premultiplied by $\mathbf{L}$ in (5.14) is given by equation (5.11).

5. Set up the state and observation equations for an AR(1) process as indicated in equations (5.15) and write out the steps for the first two observations without and with observation error in the model.

6. Set up the repeated measures model in equation (1.2) in state space form keeping ϵ_{ij} in the state equation. What

change is necessary to include serial correlation in the model? Note that unbalanced designs with different numbers of observations on different subjects or missing observations are easily handled using the state space representation.

CHAPTER 6

Autoregressive Moving Average Errors

6.1 Equally spaced observations

The autoregressive moving average model, ARMA(p,q), with zero mean can be written (Box and Jenkins, 1976; Harvey, 1981)

$$\epsilon_j = \sum_{k=1}^{p} \phi_k \epsilon_{j-k} + \eta_j + \sum_{k=1}^{q} \theta_k \eta_{j-k}, \tag{6.1}$$

where the η's are uncorrelated zero mean random variables with constant variance, σ_η^2. The signs of the θ's here agree with the definition used by Harvey, but are the negative of the Box-Jenkins' θ's. For stationarity and invertibility of the moving average, the roots of

$$1 - \sum_{k=1}^{p} \phi_k z^k = 0,$$

and

$$1 + \sum_{k=1}^{q} \theta_k z^k = 0,$$

must be outside the unit circle.

There are several equivalent state space representations for ARMA processes. In my *Technometrics* paper (Jones, 1980), I used what is known as Akaike's Markovian representation (Akaike, 1973b, 1974b, 1975). Harvey (1981) uses a different but equivalent state space representation. See also Aoki (1987) for a discussion of state space representations of ARMA processes.

The minimum length of the state vector for a state space representation of an ARMA(p,q) process is

$$m = \max(p, q+1).$$

The state vector for Akaike's Markovian representation consists of the value of the process at observation j together with predictions up to $m-1$ steps. This is denoted

$$\mathbf{s}(j) = \begin{bmatrix} \epsilon(j|j) \\ \epsilon(j+1|j) \\ \vdots \\ \epsilon(j+m-1|j) \end{bmatrix},$$

where

$$\epsilon(j|j) = \epsilon_j,$$

the process at observation j, and $\epsilon(j+l|j)$ is an l step prediction. The one step prediction assumes that the ϕ's and θ's are known as well as the random inputs, η_j, up to the observation from which the predictions start, in this case observation j,

$$\epsilon(j+1|j) = \sum_{k=1}^{p} \phi_k \epsilon_{j+1-k} + \sum_{k=1}^{q} \theta_k \eta_{j+1-k}.$$

The l-step prediction uses the previous predictions in the autoregressive part,

$$\epsilon(j+l|j) = \sum_{k=1}^{l-1} \phi_k \epsilon(j+l-k|j) + \sum_{k=l}^{p} \phi_k \epsilon_{j+l-k} + \sum_{k=l}^{q} \theta_k \eta_{j+l-k},$$

where terms in the summations are eliminated if the indices are not in the proper range.

In order to develop the state transition from observation $j-1$ to observation j, it is helpful to first consider the one-sided moving average representation of the process,

$$\epsilon_j = \eta_j + \sum_{k=1}^{\infty} \psi_k \eta_{j-k}, \tag{6.2}$$

where ψ's are the ψ weights used by Box and Jenkins (1976, p. 46). Let $\psi_0 = 1$ and $\psi_k = 0$ for $k < 0$. The ψ's are also called the impulse response function since, if all the η_j are set equal to zero except for $\eta_0 = 1$, the process output would be $\epsilon_j = \psi_j$. The recursion for calculating the ψ's from the ϕ's and θ's is (Box and Jenkins, 1976, p. 134),

$$\psi_0 = 1$$

$$\psi_l = \theta_l + \sum_{k=1}^{l} \phi_k \psi_{l-k},$$

where $\theta_l = 0$ for $l > q$ and $\phi_k = 0$ for $k > p$.

The one-sided moving average representation is useful for deriving the updating of the state vector, and for determining prediction variances and covariances. The one-sided representation of the process at observation $j+l$ is (6.2) shifted by l time units,

$$\epsilon_{j+l} = \sum_{k=0}^{\infty} \psi_k \eta_{j+l-k}.$$

Assuming that the ψ's and η's are known, an l step prediction based on the one-sided moving average, starting from observation j, is

$$\epsilon(j+l|j) = \sum_{k=l}^{\infty} \psi_k \eta_{j+l-k}.$$

This equation uses only the η's up to the time that the prediction is made. The error in this prediction is due to the η's between the time that the prediction is made and the time of the prediction,

$$\epsilon_{j+l} - \epsilon(j+l|j) = \sum_{k=0}^{l-1} \psi_k \eta_{j+l-k}. \tag{6.3}$$

The prediction of observation $j+l$ starting from observation $j-1$ is

$$\epsilon(j+l|j-1) = \sum_{k=l+1}^{\infty} \psi_k \eta_{j+l-k},$$

so that

$$\epsilon(j+l|j) = \epsilon(j+l|j-1) + \psi_l \eta_j.$$

This is the equation given in Box-Jenkins for using the ψ weights in updating forecasts.

The state transition equation for an ARMA(p,q) process is,

$$\begin{bmatrix} \epsilon(j|j) \\ \epsilon(j+1|j) \\ \vdots \\ \epsilon(j+m-1|j) \end{bmatrix} = \begin{bmatrix} 0 & 1 & 0 & \cdots & 0 \\ 0 & 0 & 1 & \cdots & 0 \\ & & \vdots & & \\ \phi_m & & \cdots & \phi_2 & \phi_1 \end{bmatrix}$$

$$\times \begin{bmatrix} \epsilon(j-1|j-1) \\ \epsilon(j|j-1) \\ \vdots \\ \epsilon(j+m-2|j-1) \end{bmatrix} + \begin{bmatrix} 1 \\ \psi_1 \\ \vdots \\ \psi_{m-1} \end{bmatrix} \eta_j. \qquad (6.4)$$

An element of the state vector at observation j consists of the element of the state vector one element lower at observation $j-1$ plus a constant multiplied by η_j. This does not work for the final element of the state vector since there is no element to shift up. The final element of the state vector can be written directly as a prediction,

$$\epsilon(j+m-1|j) = \sum_{k=1}^{p} \phi_k \epsilon(j+m-1-k|j-1) + \psi_{m-1}\eta_j.$$

This is a prediction of length m, so there is no contribution from the moving average.

The observation equation for an ARMA(p,q) process is

$$\xi(j) = \begin{bmatrix} 1 & 0 & \cdots & 0 \end{bmatrix} \begin{bmatrix} \epsilon(j|j) \\ \epsilon(j+1|j) \\ \vdots \\ \epsilon(j+m-1|j) \end{bmatrix} + v(j). \qquad (6.5)$$

As is shown by Box and Jenkins (1976, p. 122), observational error can only be included in the model if $q < p$ since this results in an ARMA process of order (p, p). If $q \geq p$ and observational error is included in the model, the result is an ARMA process of the same order, (p, q). In this case, the observational error simply changes the values of the moving average coefficients, the θ's. These q moving average coefficients and the observational error variance cannot all be estimated since they are confounded. On the other hand, if $q < p-1$, it may be more parsimonious to fit an ARMA(p, q) process with observational error than to fit an ARMA(p, p) model.

6.1.1 *The initial state covariance matrix*

In order to use the Kalman filter to calculate the exact value of -2 ln likelihood for given values of the ARMA coefficients, with or without missing observations, it is necessary to calculate the proper initial state covariance matrix. Since the random inputs to the model have zero means, the initial state vector is a vector of zeros, since this is the best estimate of the state vector before any observations are taken.

Multiplying equation (6.3) by ϵ_j, which is the same as $\epsilon(j|j)$, and taking expected values gives

$$C(l) = E(\epsilon_j \epsilon_{j+l}) = E[\epsilon(j|j)\epsilon(j+l)], \tag{6.6}$$

since the η's are uncorrelated with past ϵ's. Here $C(l)$ is the covariance function of the ARMA process at lag l. The first row and column of the initial state covariance matrix will be equal to the covariance function of the process from lag 0 to lag $m-1$. Multiplying

$$\epsilon_{j+l} = \epsilon(j+l|j) + \sum_{k=0}^{l-1} \psi_k \eta_{j+l-k},$$

by

$$\epsilon_{j+i} = \epsilon(j+i|j) + \sum_{k=0}^{i-1} \psi_k \eta_{j+i-k},$$

for $l \geq i > 0$, and taking expected values gives

$$C(l-i) = E(\epsilon_{j+i}\epsilon_{j+l}) = E[\epsilon(j+i|j)\epsilon(j+l|j)] + \sigma_\eta^2 \sum_{k=0}^{i-1} \psi_k \psi_{k+l-i},$$

since the predictions are uncorrelated with the prediction errors. These elements of the initial state covariance matrix are

$$P_{il}(0|0) = C(l-i) - \sigma_\eta^2 \sum_{k=0}^{i-1} \psi_k \psi_{k+l-i}. \tag{6.7}$$

In order to calculate the initial state covariance matrix, it is necessary to first calculate the covariance function of the process as a function of the ϕ's, θ's, and σ_η^2. The covariances satisfy

$$\begin{aligned} C(l) &= E(\epsilon_{j+l}\epsilon_j) = E\left\{\left(\sum_{k=1}^{p} \phi_k \epsilon_{j+l-k} + \sum_{k=0}^{q} \theta_k \eta_{j+l-k}\right)\epsilon_j\right\} \\ &= \sum_{k=1}^{p} \phi_k C(l-k) + \sum_{k=l}^{q} \theta_k E(\eta_{j+l-k}\epsilon_j), \end{aligned}$$

where $\theta_0 = 1$. $E(\eta_{j+l-k}\epsilon_j)$ is the cross covariance function between the random input process, η_j and the ϵ_j process at lag $k - l$. Using (6.2) this is

$$E\left\{\eta_{j+l-k}\left(\sum_{m=0}^{\infty}\psi_m\eta_{j-m}\right)\right\} = \sigma_\eta^2\psi_{k-l}.$$

The ψ weights are the cross correlation function between the random input process and output process. This gives

$$C(l) = \sum_{k=1}^{p}\phi_k C(l-k) + \sigma_\eta^2\sum_{k=l}^{q}\theta_k\psi_{k-l}.$$

These equations, for $l = 0$ to p, are

$$\begin{aligned} C(0) &= \phi_1 C(1) + \phi_2 C(2) + \cdots + \phi_p C(p) \\ &\quad + \sigma_\eta^2(\psi_0 + \theta_1\psi_1 + \cdots + \theta_q\psi_q) \\ C(1) &= \phi_1 C(0) + \phi_2 C(1) + \cdots + \phi_p C(p-1) \\ &\quad + \sigma_\eta^2(\theta_1\psi_0 + \theta_2\psi_1 + \cdots + \theta_q\psi_{q-1}) \\ &\vdots \\ C(p) &= \phi_1 C(p-1) + \phi_2 C(p-2) + \cdots + \phi_p C(0) \\ &\quad + \sigma_\eta^2(\theta_p\psi_0 + \theta_{p+1}\psi_1 + \cdots + \theta_q\psi_{q-p}). \end{aligned} \tag{6.8}$$

If $q < p$, the terms in the sums involving the ψ's will only enter up to the equation for $C(p - q)$. These equations can be rearranged to give $p + 1$ equations to solve for $C(0)$, $C(1)$, $\cdots$, $C(p)$ except for a multiplicative constant.

This again brings up the question of normalizing the variances so that one variance can be concentrated out of the likelihood. One variance can be set equal to 1 and estimated by the mean square error (MSE). The remaining variances and covariances are scaled by dividing by the variance that is concentrated out of the likelihood. At the end of the estimation procedure, all estimated variances and covariances are multiplied by the mean square error to scale them back. The two natural choices for concentrating out of the likelihood are σ_η^2 or $\sigma^2 = C(0)$. The first is the variance of the random input, and the second is the variance of the within subject error (excluding any observational error which may be included in the model). The choice used in this book is to concentrate $C(0)$ out of the likelihood by scaling all variances and covariances so that $C(0) = 1$. To do this, the equations (6.8) are solved by first assuming that $\sigma_\eta^2 = 1$. Equations (6.6) and (6.7) can then be used to calculate the unnormalized initial state covariance matrix. The value of $C(0)$ obtained from these calculations is divided into the elements of the initial state covariance matrix giving the correlation function of the process as the first row and column. The value of $C(0)$ used for this normalization is not the actual variance of the process, but its value should be saved since it is the ratio of the variance of the process to the variance of the random input. When the actual variance of the process is estimated as the mean square error (MSE), the variance of the random input is estimated as the mean square error divided by this saved ratio. Programs for these calculations are given in the Appendix on page 193. It is helpful, at this stage, to look at a simple special case.

The AR(1) case

Equations (6.8) for an AR(1) process are

$$\begin{aligned} C(0) &= \phi_1 C(1) + \sigma_\eta^2 \\ C(1) &= \phi_1 C(0). \end{aligned}$$

The rearranged system of equations is

$$\begin{bmatrix} 1 & -\phi_1 \\ \phi_1 & 1 \end{bmatrix} \begin{bmatrix} C(0) \\ C(1) \end{bmatrix} = \begin{bmatrix} \sigma_\eta^2 \\ 0 \end{bmatrix}.$$

The solution of these equations is

$$\begin{bmatrix} C(0) \\ C(1) \end{bmatrix} = \frac{\sigma_\eta^2}{1-\phi_1^2} \begin{bmatrix} 1 \\ \phi_1 \end{bmatrix}.$$

The initial state covariance matrix is a scalar in this case,

$$P(0|0) = C(0) = \sigma^2 = \text{var}(\epsilon_j) = \frac{\sigma_\eta^2}{1-\phi_1^2}.$$

Since σ^2, σ_η^2 and ϕ_1^2 are unknown, we set

$$P(0|0) = 1,$$

and estimate ϕ_1 using nonlinear optimization. When the maximum likelihood estimate of ϕ_1 is obtained, the mean square error estimates σ^2,

$$\hat{\sigma}^2 = MSE,$$

and the variance of the random input is estimate as

$$\hat{\sigma}_\eta^2 = \hat{\sigma}^2(1 - \phi_1)^2.$$

6.1.2 *Reparameterizing for stability*

Unconstrained nonlinear optimization programs attempt to minimize a function by varying the parameters over the whole real line. Autoregression coefficients of a stationary process can not take arbitrary values. For example, for an AR(1) process, ϕ_1 must be between -1 and 1. Box and Jenkins (1976, p. 58) discuss the restrictions on the autoregression coefficients for an AR(2) process. The parameter space for a stationary process becomes complicated for higher order processes.

One possibility is to parameterize the process using the partial autoregression coefficients (Box and Jenkins, 1976, p. 64). If ϕ_{kj} denotes the jth autoregression coefficient when fitting an autoregression of order k, the last coefficient, ϕ_{kk} is the kth partial autoregression coefficient. A necessary and sufficient condition for an autoregression of order p to be stationary is for all the partial autoregression coefficients from 1 to p to be between -1 and 1,

$$-1 < \phi_{kk} < 1.$$

The partial autoregression coefficients can be constrained to be in the interval $(-1, 1)$ using the logistic transformation that was used for an AR(1) coefficient in section 4.3,

$$u_k = \ln\left(\frac{1 + \phi_{kk}}{1 - \phi_{kk}}\right),$$

with the inverse transformation

$$\phi_{kk} = \frac{1 - \exp(-u_k)}{1 + \exp(-u_k)}.$$

Optimization is carried out with respect to the u_k's. Within the subroutine that calculates the likelihood, the u_k's are transformed to the ϕ_{kk}, and it is then necessary to calculate the actual autoregression coefficients from the partial autoregression coefficients.

The autoregression coefficients can be calculated from the partial autoregression coefficients using the Levinson (1947)–Durbin (1960) recursion, for $k = 2, \cdots, p$

$$\phi_{kj} = \phi_{k-1,j} - \phi_{kk}\phi_{k-1,k-j} \quad \text{for} \quad j = 1, 2, \cdots, k-1.$$

A similar transformation can be used for the moving average coefficients to constrain the estimates to be invertible. A program for untransforming the autoregression coefficients using this algorithm is given on page 189.

6.2 Unequally spaced observations

Continuous time models beyond the AR(1) model are significantly more complicated. Pandit and Wu (1983) discuss the sampling of continuous time processes at equally spaced intervals. If a continuous time ARMA(p,q) process with $q < p$ is sampled at equally spaced time intervals, the result is a discrete time ARMA(p,p–1) process. If a continuous time ARMA(p,q) process with $q < p$ and observational error is sampled at equally spaced time intervals, the result is a discrete time ARMA(p,p) process. It is therefore possible to fit a continuous time process to equally spaced observations in the hope of finding a more parsimonious model. For example, a continuous time AR(p) model would have p parameters, while the sampled process would be ARMA(p,p–1) with $2p-1$ parameters. These $2p-1$ parameters would be complicated functions of the original p continuous time autoregression coefficients. Most of the results in this section can be found in Jones and Ackerson (1990). They are repeated here for completeness. Doob (1953) discusses continuous time models in Chapter 11, section 10 (p. 542), and it is interesting that the necessary mathematics existed at that time. A continuous time ARMA(p,q) process is defined by the stochastic linear differential equation

$$\frac{d^p\epsilon(t)}{dt^p} + \alpha_{p-1}\frac{d^{p-1}\epsilon(t)}{dt^{p-1}} + \cdots + \alpha_0\epsilon(t) =$$
$$\delta_q\frac{d^q\eta(t)}{dt^q} + \delta_{q-1}\frac{d^{q-1}\eta(t)}{dt^{q-1}} + \cdots + \eta(t).$$

Here $\eta(t)$ is continuous time 'white noise', as in the AR(1) special case, which integrates to a Wiener process with variance σ_η^2 per unit time. For stationarity, it is necessary that $q < p$ and that the roots of

$$A(z) = \sum_{k=0}^{p} \alpha_k z^k = 0 \tag{6.9}$$

have negative real parts, where $\alpha_p = 1$. Linear differential equations have solutions of exponential form, and if the roots have negative real parts, disturbances to the system will die away with time. Equation (6.9) is referred to as the characteristic equation of the system. The system is 'minimum phase' or 'minimum delay' if the roots of

$$\sum_{k=0}^{q} \delta_k z^k = 0$$

have non-positive real parts, where $\delta_0 = 1$.

To avoid solving for the roots of the characteristic polynomial and to constrain the ARMA model to be stationary, the autoregression coefficients can be reparameterized as in Jones (1981). The characteristic polynomial (6.9) is written as quadratic factors with a linear factor if p is odd,

$$\begin{aligned} A(z) &= (a_1 + a_2 z + z^2)(a_3 + a_4 z + z^2)\cdots \\ &\quad \begin{cases} (a_{p-1} + a_p z + z^2) & \text{if } p \text{ is even} \\ (a_p + z) & \text{if } p \text{ is odd.} \end{cases} \end{aligned} \tag{6.10}$$

The roots will have negative real parts if and only if the a's are positive. A log transformation is used to constrain the a's to be positive, and the optimization is carried out with respect to the logs of the a's. From the a's the roots of the characteristic polynomial are easily calculated from the quadratic and linear factors. A program for calculating the roots from the logs of the a's is given on page 197.

A similar transformation can be used for the moving average part. Let

$$D(z) = \sum_{k=0}^{q} \delta_k z^k.$$

This polynomial can be factored as

$$\begin{aligned} D(z) &= (1 + d_1 z + d_2 z^2)(1 + d_3 z + d_4 z^2)\cdots \\ &\quad \begin{cases} (1 + d_{q-1} z + d_q z^2) & \text{if } q \text{ is even} \\ (1 + d_q z) & \text{if } q \text{ is odd.} \end{cases} \end{aligned} \tag{6.11}$$

A square root transformation can be used to constrain the d's to be nonnegative. For the equations that follow, it is necessary to calculate the δ's from the d's. A program to calculate the δ's from the square roots of the d's is given on page 198.

For a stationary continuous time ARMA process, the spectral density (Box and Jenkins, p. 41) is

$$s(f) = \sigma_\eta^2 \frac{|\sum_{k=0}^{q} \delta_k (2\pi i f)^k|^2}{|\sum_{k=0}^{p} \alpha_k (2\pi i f)^k|^2}$$

which is referred to as a rational spectrum (Doob, 1953). In this equation, f denotes frequency in cycles per unit time, and $i = \sqrt{-1}$. The range of f is $-\infty < f < \infty$. The covariance function of the process is

$$R(\tau) = \int_{-\infty}^{\infty} s(f) \exp(2\pi i \tau f) df,$$

where τ is the time lag. This integral can be evaluated in closed form for the rational spectral density by using the method of inverting a Fourier transform by integration in the complex plane using the method of residuals (Doob, 1953). Let the roots of the characteristic equation (6.9) be $r_1, r_2 \cdots r_p$, and assume that these roots are distinct. This assumption of distinct roots will also be used later in the numerical solution using the state space approach, and is not as restrictive as it may seem. When doing a numerical search, the probability of equal roots is very near zero unless an initial guess at the unknown parameters is used which produces equal roots. In this case the program will detect the equal roots and request new starting values. The covariance function is

$$R(\tau) = \sigma_\eta^2 \sum_{k=1}^{p} \frac{(\sum_{l=0}^{q} \delta_l r_k^l)\{\sum_{l=0}^{q} \delta_l (-r_k)^l\} \exp(r_k \tau)}{-2Re(r_k) \prod_{l \neq k} (r_l - r_k)(r_l^* + r_k)} \qquad (6.12)$$

where the product in the denominator is from $l = 1$ to p excluding $l = k$, r_k^* denotes the complex conjugate of the root, and $Re(r_k)$ denotes the real part of r_k. While equation (6.12) appears complicated, it is not difficult to calculate the covariance for a given time lag. A program is given on page 200.

For an ARMA(1,0), i.e. an AR(1), process, the product term in the denominator is replaced by 1 and the covariance function is

$$R(\tau) = \sigma_\eta^2 \frac{\exp(r_1 \tau)}{-2r_1}$$

since the single root must be real and negative. This covariance function for a continuous time AR(1) process decays exponentially as does the discrete time AR(1) covariance function; however, a discrete time AR(1) process can have a negative autoregression coefficient causing the covariance to alternate positive and negative while decaying exponentially. This can not happen in continuous time. The continuous time coefficient, α_0 must be positive producing a characteristic equation $z + \alpha_0 = 0$ with root $r_1 = -\alpha_0$, so that the covariance function is positive.

The ARMA(p,q) error structure can be generalized by allowing the addition of observational error. Letting this observational error variance be $\sigma^2 \sigma_o^2$, this is added to the diagonal of the within subject error covariance matrix. Observational error is necessary in the model when a continuous time ARMA process is used if there are replicate measurements for a subject at a given time. Without observational error, replicate measurements would necessarily be identical since a prediction for a time interval of zero would have zero variance.

6.2.1 The state space method

Kalman's recursive optimal estimation procedure for state space models in discrete time was generalized by Kalman and Bucy (1961) to continuous time state space models. By integrating a continuous time state space representation over time intervals, a discrete time state space representation is developed at unequally spaced observation times.

A state space representation of a continuous time ARMA(p,q) process is (Wiberg, 1971)

$$\frac{d}{dt} \begin{bmatrix} s(t) \\ s^{(1)}(t) \\ \vdots \\ s^{(p-1)}(t) \end{bmatrix} =$$

$$\begin{bmatrix} 0 & 1 & 0 & \cdots & 0 \\ 0 & 0 & 1 & \cdots & 0 \\ & \vdots & & & \\ -\alpha_0 & -\alpha_1 & -\alpha_2 & \cdots & -\alpha_{p-1} \end{bmatrix} \begin{bmatrix} s(t) \\ s^{(1)}(t) \\ \vdots \\ s^{(p-1)}(t) \end{bmatrix} + \begin{bmatrix} 0 \\ 0 \\ \vdots \\ 1 \end{bmatrix} \eta(t), \tag{6.13}$$

and the observation equation for an observation at time t_j is

$$\epsilon(t_j) = \begin{bmatrix} 1 & \delta_1 & \cdots & \delta_{p-1} \end{bmatrix} \begin{bmatrix} s(t_j) \\ s^{(1)}(t_j) \\ \vdots \\ s^{(p-1)}(t_j) \end{bmatrix} + v(t_j),$$

where $v(t_j)$ is the observational error, if present, with variance $\sigma^2\sigma_o^2$. The k'th derivative of $s(t)$ with respect to t is denoted by $s^{(k)}$, and $\delta_k = 0$ if $k > q$. In matrix notation, the state space representation is

$$\frac{d}{dt}\mathbf{s}(t) = \mathbf{F}\mathbf{s}(t) + \mathbf{g}\eta(t)$$

$$\epsilon(t_j) = \mathbf{h}\mathbf{s}(t_j) + v(t_j), \tag{6.14}$$

These equations express the elements of the within subject error vector ϵ_i in the longitudinal data model (2.1) in terms of a state vector $\mathbf{s}(t)$ which is a continuous time random process and $p-1$ of its derivatives.

In order to put equation (2.1) in state space form it is necessary to augment the state vector by the vector γ_i. The complete state space representation for subject i is

$$\frac{d}{dt}\begin{bmatrix} \mathbf{s}_i(t) \\ \gamma_i \end{bmatrix} \begin{bmatrix} \mathbf{F} & 0 \\ 0 & 0 \end{bmatrix} \begin{bmatrix} \mathbf{s}_i(t) \\ \gamma_i \end{bmatrix} + \begin{bmatrix} \mathbf{g} \\ 0 \end{bmatrix} \eta_i(t) \tag{6.15}$$

$$\xi_i(t_j) = \begin{bmatrix} \mathbf{h} & \mathbf{z}(t_j) \end{bmatrix} \begin{bmatrix} \mathbf{s}_i(t_j) \\ \gamma_i \end{bmatrix} + v_i(t_j). \tag{6.16}$$

Here $\mathbf{z}(t_j)$ is a row vector consisting of row j in $\mathbf{Z}_i$ of equation (2.1). Equation (6.15) indicates that the random effects for subject i are constant over time. The fact that the random effects vary from subject to subject is introduced into the model through the initial state covariance matrix which will be set equal to $B = \mathbf{U}'\mathbf{U}$. Since the elements of this prior covariance matrix will be estimated by maximum likelihood, this is an empirical Bayes method.

Jones (1981) used the state space approach to calculate the likelihood for a continuous time stationary AR(p) time series with zero mean. This is the case shown in equations (6.13) to (6.14) without the δ's in the **h** vector. The method consists of integrating the state equation over the intervals between observations to obtain a discrete state space representation corresponding to the observation points. Since the δ's appear in the observation equation, this method applies to the present situation. Integrating the state equation, with the random input removed, over a time interval from t_{j-1} to t_j gives the matrix exponential solution,

$$\mathbf{s}(t_j) = \exp\{(t_j - t_{j-1})\mathbf{F}\}\mathbf{s}(t_{j-1}). \tag{6.17}$$

There are many possible ways to evaluate matrix exponentials numerically (Moler & Van Loan, 1978). Because of the structure of the **F** matrix, the eigenvalues are the roots of the characteristic equation (6.9), r_k, and assuming that these roots are distinct, the matrix **F** can be rotated to diagonal form. The right eigenvector corresponding to the root r_k has l'th element

$$C_{lk} = r_k^{l-1}. \tag{6.18}$$

The diagonal form of **F** is

$$\mathbf{F} = \mathbf{C}\mathbf{\Lambda}\mathbf{C}^{-1},$$

where **C** is a square matrix, the columns of which are the right eigenvectors of **F**, and $\mathbf{\Lambda}$ is a diagonal matrix with diagonal elements r_k. These matrices may contain complex elements so complex arithmetic is necessary for these calculations. Equation (6.17) can be written

$$\mathbf{s}(t_j) = \mathbf{C}\exp\{(t_j - t_{j-1})\mathbf{\Lambda}\}\mathbf{C}^{-1}\mathbf{s}(t_{j-1}).$$

Defining a complex rotated state vector

$$\tilde{\mathbf{s}}(t) = \mathbf{C}^{-1}\mathbf{s}(t)$$

produces an uncoupled system where each element of the rotated state vector is predicted independently of the other elements,

$$\tilde{s}_k(t_j) = \exp\{(t_j - t_{j-1})r_k\}\tilde{s}_k(t_{j-1}),$$

or, in matrix form

$$\tilde{\mathbf{s}}(t_j) = \mathbf{\Phi}(t_j - t_{j-1})\tilde{\mathbf{s}}(t_{j-1}),$$

where

$$\mathbf{\Phi}(t_j - t_{j-1}) = \exp\{(t_j - t_{j-1})\mathbf{\Lambda}\}$$

is the diagonal, possibly complex, rotated state transition matrix. It is not necessary to actually rotate the state vector since for each subject the initial state vector is zero, so the initial rotated state vector is zero. The rotated state vector is predicted from time point to time point and need only be rotated back to obtain the actual observations using the inverse equation

$$\mathbf{s}(t) = \mathbf{C}\tilde{\mathbf{s}}(t).$$

In addition to predicting the state equation over an arbitrary time interval, it is necessary to calculate the contribution of the error term to this prediction. The complex prediction error due to $\eta(t)$ of the rotated state vector is (Kalman & Bucy, 1961)

$$\int_{t_{j-1}}^{t_j} \mathbf{\Phi}(t_j - t)\mathbf{C}^{-1}\mathbf{g}\eta(t)dt.$$

Let

$$\boldsymbol{\kappa} = \mathbf{C}^{-1}\mathbf{g} \tag{6.19}$$

which is the last column of $\mathbf{C}^{-1}$ since $\mathbf{g}$ is a column vector of zeros with a one in the last position. The complex prediction error is (Jones, 1981)

$$\int_{t_{j-1}}^{t_j} \begin{bmatrix} \kappa_1 \exp\{(t_j - t)r_1\} \\ \kappa_2 \exp\{(t_j - t)r_2\} \\ \vdots \\ \kappa_p \exp\{(t_j - t)r_p\} \end{bmatrix} \eta(t)dt,$$

with a Hermitian covariance matrix that has elements

$$\sigma_\eta^2 Q_{kl}(t_j - t_{j-1}) = \sigma_\eta^2 \frac{\kappa_k \kappa_l^*}{-(r_k + r_l^*)}[1 - \exp\{(r_k + r_l^*)(t_j - t_{j-1})\}]. \tag{6.20}$$

A Hermitian matrix is the complex analogue of a symmetric matrix. It is a matrix that is equal to its complex conjugate transposed. This produces the one step prediction covariance matrix

$\mathbf{Q}(t_j - t_{j-1})$ necessary for the Kalman filter recursion. The discrete time rotated state space representation corresponding to equation (6.15) is

$$\begin{bmatrix} \tilde{\mathbf{s}}_i(t_j) \\ \gamma_i \end{bmatrix} \begin{bmatrix} \Phi(t_j - t_{j-1}) & 0 \\ 0 & \mathbf{I} \end{bmatrix} \begin{bmatrix} \tilde{\mathbf{s}}_i(t_{j-1}) \\ \gamma_i \end{bmatrix} + \begin{bmatrix} \zeta(t_j) \\ 0 \end{bmatrix},$$

where $\zeta(t_j)$ is a zero mean complex Gaussian random vector with covariance matrix $\sigma_\eta^2\mathbf{Q}(t_j - t_{j-1})$ and independent at the observation times. The corresponding observation equation (6.16) becomes

$$\xi_i(t_j) = \begin{bmatrix} \mathbf{hC} & \mathbf{z}(t_j) \end{bmatrix} \begin{bmatrix} \tilde{\mathbf{s}}_i(t_j) \\ \gamma_i \end{bmatrix} + v_i(t_j).$$

The initial state covariance matrix for a subject is the covariance matrix of the state vector before any observations are taken on that subject. For the upper part of the state vector, this is the steady state covariance matrix of the ARMA process and its $p-1$ derivatives (Doob, 1953; Jones, 1981), and is obtained by complex integration similar to the method of obtaining the covariance function

$$\sigma_\eta^2 P_{\mu\nu}(t_1|0) = \sigma_\eta^2 \sum_{k=1}^{p} \frac{r_k^\mu(-r_k)^\nu}{-2Re(r_k)\prod_{l\neq k}(r_l - r_k)(r_l^* + r_k)}. \quad (6.21)$$

The initial covariance matrix of the rotated state vector is

$$\tilde{\mathbf{P}}_c(t_1|0) = \mathbf{C}^{-1}\mathbf{P}(t_1|0)(\mathbf{C}^*)^{-1} \quad (6.22)$$

where $\mathbf{C}^*$ denotes the complex conjugate transposed matrix. This covariance matrix may be complex and is Hermitian. The initial state covariance matrix of the complete state vector in partitioned form is

$$\tilde{\mathbf{P}}(t_1|0) = \sigma_\eta^2 \begin{bmatrix} \tilde{\mathbf{P}}_{11}(t_1|0) & \tilde{\mathbf{P}}_{12}(t_1|0) \\ \tilde{\mathbf{P}}_{12}^*(t_1|0) & \mathbf{P}_{22}(t_1|0) \end{bmatrix} = \sigma_\eta^2 \begin{bmatrix} \tilde{\mathbf{P}}_c(t_1|0) & 0 \\ 0 & \mathbf{U}'\mathbf{U} \end{bmatrix}. \quad (6.23)$$

This completes the background necessary to develop the state space approach for calculating the exact likelihood.

Since there is a state vector for each column of $\mathbf{X}_i$ and $\mathbf{y}_i$, the state is now a $(p+g)\times(b+1)$ matrix. This matrix of transformed state vectors will be denoted

$$\tilde{\mathbf{S}}(t) = \begin{bmatrix} \tilde{\mathbf{S}}_c(t) \\ \mathbf{S}_r(t) \end{bmatrix}.$$

Here, $\tilde{\mathbf{S}}_c(t)$ denotes the first part of the state corresponding to the ARMA process and the subscript c indicates that this is complex. The subscript of the second part of the state vector, which is the part corresponding to the random effects is r, which also indicates that this part of the state is real. The subscript i denoting the subject has been suppressed. The state matrix is initialized to zero for each subject with its initial covariance matrix set to $\tilde{\mathbf{P}}(t_1|0)$, (6.23).

1. The state matrix is predicted to the next observation time,

$$\begin{bmatrix} \tilde{\mathbf{S}}_c(t_j|t_{j-1}) \\ \mathbf{S}_r(t_j|t_{j-1}) \end{bmatrix} = \begin{bmatrix} \mathbf{\Phi}(t_j - t_{j-1}) & 0 \\ 0 & \mathbf{I} \end{bmatrix} \begin{bmatrix} \tilde{\mathbf{S}}_c(t_{j-1}|t_{j-1}) \\ \mathbf{S}_r(t_{j-1}|t_{j-1}) \end{bmatrix}.$$

   The state transition matrix in this equation is diagonal.

2. The covariance matrix of this prediction is

$$\begin{aligned} \tilde{\mathbf{P}}_{11}(t_j|t_{j-1}) &= \mathbf{\Phi}(t_j - t_{j-1})\tilde{\mathbf{P}}_{11}(t_{j-1}|t_{j-1})\mathbf{\Phi}^*(t_j - t_{j-1}) \\ &+ \mathbf{Q}(t_j - t_{j-1}) \\ \tilde{\mathbf{P}}_{12}(t_j|t_{j-1}) &= \mathbf{\Phi}(t_j - t_{j-1})\tilde{\mathbf{P}}_{12}(t_{j-1}|t_{j-1}) \\ \mathbf{P}_{22}(t_j|t_{j-1}) &= \mathbf{P}_{22}(t_{j-1}|t_{j-1}) \end{aligned}$$

   These first two steps are skipped at the first observation for each subject. $\mathbf{P}_{11}$ and $\mathbf{P}_{12}$ may be complex, but $\mathbf{P}_{22}$ will remain real.

3. The innovation (a real row vector of length $b+1$) at time t_j for subject i is

$$\mathbf{I}_i(t_j) = \begin{bmatrix} \mathbf{x}_i(t_j) & y_i(t_j) \end{bmatrix} - \begin{bmatrix} \mathbf{hC} & \mathbf{z}(t_j) \end{bmatrix} \begin{bmatrix} \tilde{\mathbf{S}}_c(t_j|t_{j-1}) \\ \mathbf{S}_r(t_j|t_{j-1}) \end{bmatrix},$$

   where $\mathbf{x}_i(t_j)$ is a row vector consisting of row j in $\mathbf{X}_i$ of (2.1), and $y_i(t_j)$ is the corresponding element of $\mathbf{y}_i$.

4. The innovation variance is a scalar,

$$\begin{aligned} V_i(t_j) &= (\mathbf{hC})\tilde{\mathbf{P}}_{11}(t_j|t_{j-1})(\mathbf{hC})^* + \\ & 2Re\{(\mathbf{hC})\tilde{\mathbf{P}}_{12}(t_j|t_{j-1})\mathbf{z}'(t_j)\} + \\ & \mathbf{z}(t_j)\mathbf{P}_{22}(t_j|t_{j-1})\mathbf{z}'(t_j) + \sigma_o^2. \end{aligned}$$

5. The upper triangular part of a $(b+1)\times(b+1)$ matrix necessary for calculating the estimates of β and σ^2, and the determinant term are accumulated over all the observations for all the subjects,

$$\mathbf{M} \leftarrow \mathbf{M} + \mathbf{I}_i'(t_j)\mathbf{I}_i(t_j)/V_i(t_j),$$

$$\Delta \leftarrow \Delta + \log\{V_i(t_j)\}.$$

where $\leftarrow$ indicates that the left side is replaced by the right side of the expression.

6. The Kalman gain is

$$\mathbf{K}_i(t_j) = \frac{1}{V_i(t_j)}\begin{bmatrix} \tilde{\mathbf{P}}_{11}(t_j|t_{j-1})(\mathbf{hC})^* + \tilde{\mathbf{P}}_{12}(t_j|t_{j-1})\mathbf{z}'(t_j) \\ \{(\mathbf{hC})\tilde{\mathbf{P}}_{12}(t_j|t_{j-1})\}^* + \mathbf{P}_{22}(t_j|t_{j-1})\mathbf{z}'(t_j) \end{bmatrix}.$$

7. The updated estimate of the state is

$$\begin{bmatrix} \tilde{\mathbf{S}}_c(t_j|t_j) \\ \mathbf{S}_r(t_j|t_j) \end{bmatrix} = \begin{bmatrix} \tilde{\mathbf{S}}_c(t_j|t_{j-1}) \\ \mathbf{S}_r(t_j|t_{j-1}) \end{bmatrix} + \mathbf{K}_i(t_j)\mathbf{I}_i(t_j).$$

8. The updated state covariance matrix is

$$\begin{bmatrix} \tilde{\mathbf{P}}_{11}(t_j|t_j) & \tilde{\mathbf{P}}_{12}(t_j|t_j) \\ \tilde{\mathbf{P}}_{12}^*(t_j|t_j) & \mathbf{P}_{22}(t_j|t_j) \end{bmatrix} =$$

$$\begin{bmatrix} \tilde{\mathbf{P}}_{11}(t_j|t_{j-1}) & \tilde{\mathbf{P}}_{12}(t_j|t_{j-1}) \\ \tilde{\mathbf{P}}_{12}^*(t_j|t_{j-1}) & \mathbf{P}_{22}(t_j|t_{j-1}) \end{bmatrix} - V_i(t_j)\mathbf{K}_i(t_j)\mathbf{K}_i^*(t_j).$$

This completes the recursion. Because of the subtraction in this equation, the recursion for the covariance matrix would usually be carried out in double precision.

The state and state covariance matrix are reinitialized for each subject, M and Δ are accumulated over all the observations for all the subjects, and -2 log likelihood is evaluated at the end. The matrix M is the same matrix generated by the direct method in equation (2.17) with the total sum of squares (TSS) in the lower right hand position, and -2 log likelihood is calculated from this point as in the direct method.

6.3 An exercise

1. What is the initial covariance matrix for an equally spaced ARMA(1,1) process. Start by writing out this special case of equations (6.8).

CHAPTER 7

Nonlinear Models

There are many examples of nonlinear models in the biological literature. An example is the Michaelis-Menten pharmacokinetics model which approaches a horizontal asymptote caused by saturation in the reaction (see, for example, Ruppert, Cressie and Carroll, 1989). The simplest form of the model is

$$y_j = \frac{x_j V}{x_j + K} + \epsilon_j, \tag{7.1}$$

where y_j is the response at concentration $x_j \geq 0$. This equation is linear in V which is the asymptote, and nonlinear with respect to K which is called the Michaelis constant, and is the value of x where the height of the curve is at half the upper asymptote, V. Fitting this type of model to a single curve is a standard application of nonlinear least squares (Draper and Smith, 1981).

7.1 Fitting nonlinear models

Assume that the errors in model (7.1) are independent Gaussian with zero means and constant variance σ^2. Since both parameters in the model, V and K, must be positive, the variable transformations

$$\begin{aligned} \beta_1 &= \ln V \\ \beta_2 &= \ln K \end{aligned}$$

will ensure that V and K will remain positive for any values of β_1 and β_2 that the optimization program may try. Equation (7.1) can be written

$$y_j = \frac{x_j \exp(\beta_1)}{x_j + \exp(\beta_2)} + \epsilon_j, \tag{7.2}$$

The maximum likelihood solution is then the values of the β's that minimize the residual sum of squares,

$$RSS = \sum_{j=1}^{n} \left[y_j - \frac{x_j \exp(\hat{\beta}_1)}{x_j + \exp(\hat{\beta}_2)} \right]^2 . \tag{7.3}$$

Since the model is linear in $\exp(\beta_1)$, for a given value of $\hat{\beta}_2$, the maximum likelihood estimate of $\exp(\beta_1)$ is

$$\exp(\hat{\beta}_1) = \frac{\sum_{j=1}^{n} \left(\frac{x_j}{x_j + \exp(\hat{\beta}_2)} \right) y_j}{\sum_{j=1}^{n} \left(\frac{x_j}{x_j + \exp(\hat{\beta}_2)} \right)^2} . \tag{7.4}$$

This is the usual equation for estimating the slope when fitting a straight line through the origin with x_j replaced by $x_j/[x_j + \exp(\hat{\beta}_2)]$. Since maximum likelihood estimates are invariant under transformations of the parameters, the maximum likelihood estimate of β_1, for a given value of $\hat{\beta}_2$, is

$$\hat{\beta}_1 = \ln \left[\frac{\sum_{j=1}^{n} \left(\frac{x_j}{x_j + \exp(\hat{\beta}_2)} \right) y_j}{\sum_{j=1}^{n} \left(\frac{x_j}{x_j + \exp(\hat{\beta}_2)} \right)^2} \right] .$$

Substituting (7.4) into (7.3) gives the residual sum of squares as a function of $\hat{\beta}_2$, $RSS(\hat{\beta}_2)$. It is now possible to search for a minimum residual sum of squares by writing an interactive program where $\hat{\beta}_2$ is input and $RSS(\hat{\beta}_2)$ is output. Even better, it is possible to plot $RSS(\hat{\beta}_2)$ as a function of $\hat{\beta}_2$. This not only helps locate the minimum but can be used to obtain approximate confidence limits on $\hat{\beta}_2$. Jones and Molitoris (1984) used this method to obtain approximate confidence intervals on the 'breakpoint' or 'change point' of a line. The mean square error is estimated from the residual sum of squares at the minimum as

$$MSE = RSS_{min}/(n-2),$$

since two parameters are being estimated. As $\hat{\beta}_2$ moves away from the minimum, RSS, of course, increases. Moving away from the minimum to the point where $\hat{\beta}_2$ just becomes significantly different from the value at the minimum gives the approximate confidence interval, and this interval need not be symmetric. This point can be determined using the 'extra sum of squares' principal (Draper and Smith, 1981, p. 97),

$$\frac{RSS - RSS_{min}}{MSE} \sim F_{1,n-2},$$

where $F_{1,n-2}$ is a given significance level obtained from the F distribution with 1 and $n-2$ degrees of freedom. For example, if n is large and the 5% level is used, the ends of the confidence intervals will be where the RSS reaches

$$RSS = RSS_{min} + 3.84\, MSE.$$

Using the linearization or Taylor series method (Draper and Smith, 1981, p. 462), the nonlinear function is expanded in a Taylor series about an initial guess at the parameters. For a general equation

$$y_j = f(x_j, \beta) + \epsilon_j,$$

the expansion about the initial guesses $\beta^{(0)}$ is

$$y_j \sim f(x_j, \beta_k^{(0)}) + \sum_k \frac{\partial f}{\partial \beta_k}\left(\beta_k - \beta_k^{(0)}\right) + \epsilon_j.$$

For equation (7.2), this is

$$\begin{aligned} y_j \quad \sim \quad & \frac{x_j \exp(\beta_1^{(0)})}{x_j + \exp(\beta_2^{(0)})} + \frac{x_j \exp(\beta_1^{(0)})}{x_j + \exp(\beta_2^{(0)})}\left(\beta_1 - \beta_1^{(0)}\right) \\ - \quad & \frac{x_j \exp(\beta_1^{(0)} + \beta_2^{(0)})}{[x_j + \exp(\beta_2^{(0)})]^2}\left(\beta_2 - \beta_2^{(0)}\right) + \epsilon_j. \end{aligned} \tag{7.5}$$

Using this equation in an iterative scheme, the $\beta^{(0)}$'s represent the estimates from the previous iteration rather than the initial guesses. Moving the first term on the right to the left hand side of the equations gives the residuals from the previous iteration on the left,

$$e_j = y_j - \frac{x_j \exp(\beta_1^{(0)})}{x_j + \exp(\beta_2^{(0)})}.$$

The factors $(\beta_k - \beta_k^{(0)})$ are corrections to the estimates at the previous iterations. The corrections are now in the form of a linear equation with the residuals from the previous iteration on the left hand side and the partial derivatives of the nonlinear function with respect to the parameters as the independent variables. These partial derivatives will change at each iteration as the values of the $\beta_k^{(0)}$ change. Solving the equations by least squares for the corrections to the β's at the previous iteration will often converge to the least squares solution, especially if the initial guesses are good. When an estimated correction actually increases the residual sum of squares rather than decrease it, the problem may be that the step is too big and we crossed over a valley and up the other side of the hill.

The field of nonlinear optimization is highly developed and there are many experts in this field. The applied statistician does not need to become involved with the deeper aspects of optimization. Routines exist where it is only necessary to calculate the function that is to be minimized for given values of the unknown parameters. Some require that gradients (derivatives of the function with respect to the unknown parameters) be supplied and some do not. To get approximate asymptotic covariance matrices of the estimated nonlinear parameters, Fisher's information matrix can be approximated numerically (Dennis and Schnabel, 1983) at the minimum of -2 ln likelihood. This is the Hessian or second derivative matrix of $-$ln likelihood (beware of the factor of two if -2 ln likelihood is being minimized).

The linearization method may not be the best method for finding nonlinear estimates because of non-convergence problems. One popular method for improving the convergence properties is Marquardt's compromise (Draper and Smith, 1981, p 471). This method increases the diagonal elements of the $\mathbf{X}'\mathbf{X}$ of the linearized normal equations which rotates the step of the correction vector of the regression coefficients to a more downhill direction, and shortens the step length. This means that it is not as likely to go over a minimum and up the other side.

7.2 Nonlinear random effects

7.2.1 The Michaelis-Menten equation

When there are groups of curves, assuming that the coefficients are random across curves causes problems. A random effects version

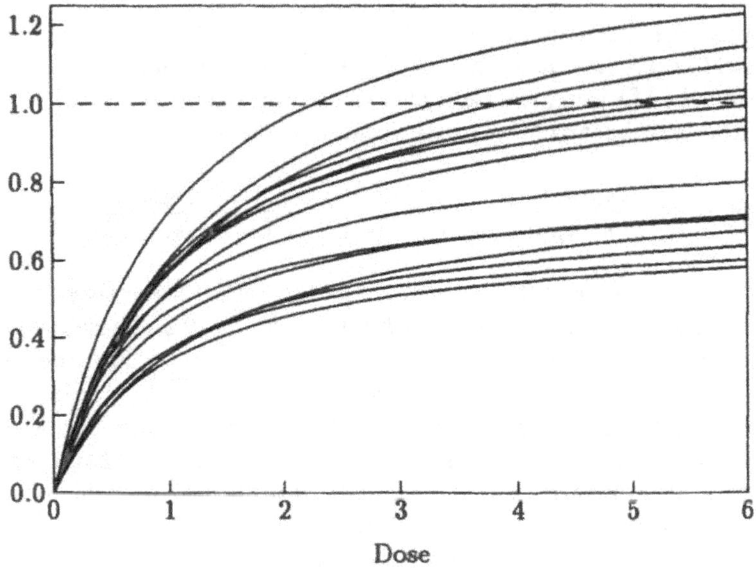

Fig. 7.1 *Simulated Michaelis-Menten curves for different subjects with random coefficients. The dashed line is the upper asymptote for a subject with* $\gamma_{i1} = 0$

of equation (7.1) is

$$y_{ij} = \frac{x_{ij} \exp(\beta_1 + \gamma_{i1})}{x_{ij} + \exp(\beta_2 + \gamma_{i2})} + \epsilon_{ij}, \tag{7.6}$$

where the subscript i denotes the curve. If we assume that the two random effects, γ_{i1} and γ_{i2} have a general zero mean bivariate Gaussian distribution across curves, the distribution of y_{ij} is no longer Gaussian because of the nonlinear transformation involving the γ's. Fig. 7.1 shows simulated curves from model (7.6) representing different subjects. The values used in the simulation were $\beta_1 = 0$, $\beta_2 = 0$, and

$$\begin{bmatrix} \gamma_{i1} \\ \gamma_{i2} \end{bmatrix} \sim N\left(\begin{bmatrix} 0 \\ 0 \end{bmatrix}, \begin{bmatrix} .04 & .02 \\ .02 & .04 \end{bmatrix} \right). \tag{7.7}$$

The two γ's have a standard deviation of 0.2, and a correlation of 0.5.

Sheiner and Beal (1980) have developed a method and a computer program, *NONMEM*, (nonlinear mixed effect model) for fitting pharmacokinetics models with random effects that appear nonlinearly (see also Hirst *et al.*, 1991). In particular, they study model (7.6) using simulations. The method used by Sheiner and Beal is to linearize (7.6) with respect to the γ's,

$$y_{ij} \sim \frac{x_{ij}\exp(\beta_1)}{x_{ij}+\exp(\beta_2)} + \frac{x_{ij}\exp(\beta_1)}{x_{ij}+\exp(\beta_2)}\gamma_{i1} - \frac{x_{ij}\exp(\beta_1+\beta_2)}{[x_{ij}+\exp(\beta_2)]^2}\gamma_{i2} + \epsilon_{ij}. \tag{7.8}$$

The coefficients of the γ's in (7.8) are the partial derivatives of the nonlinear function in (7.6) with respect to the γ's. These partial derivatives are then evaluated at $\gamma_{i1} = \gamma_{i2} = 0$, the expected values of these random variables.

From (7.8), it is possible to calculate an approximate likelihood by assuming that the γ's have a bivariate normal distribution with an arbitrary (2×2) covariance matrix, $\sigma^2\mathbf{B}$. If the γ's in (7.6), have a bivariate normal distribution, the γ's in the linearized equation, (7.8), will not have a normal distribution because of the nonlinear transformation of γ_{i2}. Since the actual distribution of the γ's is not known in a real situation, it is probably better to forget the assumption of normality in (7.6) and simply assume that the γ's in the linearized equation, (7.8) are approximately normal.

For the purpose of obtaining approximate maximum likelihood estimates, factor $\mathbf{B}$ as $\mathbf{B} = \mathbf{U}'\mathbf{U}$ where $\mathbf{U}$ is upper triangular,

$$\begin{bmatrix} b_{11} & b_{12} \\ b_{21} & b_{22} \end{bmatrix} = \begin{bmatrix} u_{11} & 0 \\ u_{12} & u_{22} \end{bmatrix} \begin{bmatrix} u_{11} & u_{12} \\ 0 & u_{22} \end{bmatrix}.$$

A nonlinear optimization program can now be used to estimate the five parameters, β_1, β_2, u_{11}, u_{12} and u_{22}. It is necessary to write a program to evaluate the approximate -2 ln likelihood as a function of these five unknown parameters. Let $\hat{\beta}_1$ and $\hat{\beta}_2$ be the current estimates of the fixed parameters, and

$$e_{ij} = y_{ij} - \frac{x_{ij}\exp(\beta_1)}{x_{ij}+\exp(\beta_2)}, \tag{7.9}$$

be the residuals. Denote the partial derivatives as,

$$z_{ij1} = \frac{x_{ij}\exp(\hat{\beta}_1)}{x_{ij}+\exp(\hat{\beta}_2)} \quad \text{and} \quad z_{ij2} = -\frac{x_{ij}\exp(\hat{\beta}_1+\hat{\beta}_2)}{[x_{ij}+\exp(\hat{\beta}_2)]^2}, \tag{7.10}$$

corresponding to the elements of the $\mathbf{Z}_i$ matrix in (2.1). In matrix notation, (7.8) can now be written,

$$\mathbf{e}_i \sim \mathbf{Z}_i\gamma_i + \epsilon_i, \qquad (7.11)$$

which is in the general form of (2.1) except that the left hand side of the equation is replaced by residuals, and the term involving fixed effects, $\mathbf{X}_i\boldsymbol{\beta}$, is deleted.

Let

$$\hat{\mathbf{U}} = \begin{bmatrix} \hat{u}_{11} & \hat{u}_{12} \\ 0 & \hat{u}_{22} \end{bmatrix},$$

where $\hat{u}_{11}$, $\hat{u}_{12}$ and $\hat{u}_{22}$ are the current estimates of u_{11}, u_{12} and u_{22}, and let

$$\mathbf{V}_i = \mathbf{Z}_i\mathbf{B}\mathbf{Z}_i' + \mathbf{W}_i = (\mathbf{Z}_i\hat{\mathbf{U}}')(\mathbf{Z}_i\hat{\mathbf{U}}')' + \mathbf{W}_i, \qquad (7.12)$$

where $\sigma^2\mathbf{W}_i$ is the covariance matrix of ϵ_i. As in (2.8) and (2.9),

$$\hat{\sigma}^2 = \frac{1}{n}\sum_i \epsilon_i'\mathbf{V}_i^{-1}\epsilon_i, \qquad (7.13)$$

and

$$\ell = n\ln(2\pi\hat{\sigma}^2) + \sum_i(\ln|\mathbf{V}_i|) + n. \qquad (7.14)$$

The function (7.14) is minimized with respect to the five parameters to obtain approximate maximum likelihood estimates.

To obtain approximate restricted maximum likelihood (REML) estimates, (7.13) is replaced by

$$\hat{\sigma}^2 = \frac{1}{n-2}\sum_i \epsilon_i'\mathbf{V}_i^{-1}\epsilon_i,$$

and (7.14) by

$$\begin{aligned} \ell &= n\ln(2\pi) + (n-2)\ln(\hat{\sigma}^2) \\ &+ \ln|\sum_i \mathbf{X}_i\mathbf{V}_i^{-1}\mathbf{X}_i| + \sum_i(\ln|\mathbf{V}_i|) + n - 2. \qquad (7.15) \end{aligned}$$

In this equation, $\mathbf{X}_i$ represents the partial derivatives of the nonlinear function with respect to the β's evaluated at $\gamma_{i1} = \gamma_{i2} = 0$. Since the random effects are added to the fixed effects, $\mathbf{X}_i = \mathbf{Z}_i$. In cases where not all of the fixed effects have additive random effects, the $\mathbf{X}_i$ matrices will have column dimension equal to the number of fixed β's, and the $\mathbf{Z}_i$ matrices will consist of the subset of these columns that have additive random effects.

Lindstrom and Bates (1990) improve on Sheiner and Beal's method by expanding the nonlinear function in (7.6) about the estimated values of the γ's for each subject rather than the mean values of zero. By expanding about the estimated values of the γ's for each subject, the linearized equation is a better approximation to the nonlinear function. Letting $\hat{\gamma}$ represent the estimated values of the γ's, the linearized equation is

$$\begin{aligned} y_{ij} &\sim \frac{x_{ij}\exp(\hat{\beta}_1+\hat{\gamma}_{i1})}{x_{ij}+\exp(\hat{\beta}_2+\hat{\gamma}_{i2})} + \frac{x_{ij}\exp(\hat{\beta}_1+\hat{\gamma}_{i1})}{x_{ij}+\exp(\hat{\beta}_2+\hat{\gamma}_{i2})}(\gamma_{i1}-\hat{\gamma}_{i1}) \\ &- \frac{x_{ij}\exp(\hat{\beta}_1+\hat{\gamma}_{i1}+\hat{\beta}_2+\hat{\gamma}_{i2})}{[x_{ij}+\exp(\hat{\beta}_2+\hat{\gamma}_{i2})]^2}(\gamma_{i2}-\hat{\gamma}_{i2})+\epsilon_{ij}. \qquad (7.16) \end{aligned}$$

The residuals, as in (7.9) are obtained by subtracting all the terms on the right hand side of (7.16) from y_{ij} except the two terms multiplying γ_{i1} and γ_{i2},

$$\begin{aligned} e_{ij} = y_{ij} &- \frac{x_{ij}\exp(\hat{\beta}_1+\hat{\gamma}_{i1})}{x_{ij}+\exp(\hat{\beta}_2+\hat{\gamma}_{i2})} + \frac{x_{ij}\exp(\hat{\beta}_1+\hat{\gamma}_{i1})}{x_{ij}+\exp(\hat{\beta}_2+\hat{\gamma}_{i2})}\hat{\gamma}_{i1} \\ &- \frac{x_{ij}\exp(\hat{\beta}_1+\hat{\gamma}_{i1}+\hat{\beta}_2+\hat{\gamma}_{i2})}{[x_{ij}+\exp(\hat{\beta}_2+\hat{\gamma}_{i2})]^2}\hat{\gamma}_{i2}+\epsilon_{ij}. \qquad (7.17) \end{aligned}$$

The z's, as in (7.10), are the terms multiplying γ_{i1} and γ_{i2},

$$z_{ij1} = \frac{x_{ij}\exp(\hat{\beta}_1+\hat{\gamma}_{i1})}{x_{ij}+\exp(\hat{\beta}_2+\hat{\gamma}_{i2})}$$

and

$$z_{ij2} = -\frac{x_{ij}\exp(\hat{\beta}_1+\hat{\gamma}_{i1}+\hat{\beta}_2+\hat{\gamma}_{i2})}{[x_{ij}+\exp(\hat{\beta}_2+\hat{\gamma}_{i2})]^2}. \qquad (7.18)$$

The value of -2 ln likelihood or the modified -2 ln likelihood is now calculated as in (7.14) or (7.15)

An updated estimate of the random effect vector for subject i is calculated using an equation similar to (2.25). Residuals are calculated with the random effects set equal to zero, i.e. using only the fixed effect,

$$e_{ij}^{(f)} = y_{ij} - \frac{x_{ij}\exp(\hat{\beta}_1)}{x_{ij}+\exp(\hat{\beta}_2)}.$$

Arranging these residuals in a vector, the updated estimate of the random effect vector is

$$\hat{\gamma}_i = \mathbf{B}\mathbf{Z}_i'\mathbf{V}_i^{-1}\mathbf{e}_i^{(f)}. \tag{7.19}$$

Given values of the β's, and using initial estimates for the random effects of zero, the steps from (7.17) to (7.19) can be iterated three or four times to get stable estimates of the random effects and values of -2 ln likelihood. The value of -2 ln likelihood is minimized with respect to the β's and u's using a nonlinear optimization routine.

7.2.2 *Simulated data*

From the simulated random effect curves shown in Fig. 7.1, observations were simulated by adding the within subject error, the ϵ_{ij}. The standard deviation of the added error was 0.1 times the value of the curve at that point. In other words, the additive error had a coefficient of variation of 10%. The simulated data are shown in Table 7.1. The actual values of the random effects used to generate the data in Table 7.1 are shown in Table 7.2.

The distribution of the γ's are as shown in (7.7). In the simulation, $\sigma^2 = 1$, and

$$\sigma\mathbf{U} = \begin{bmatrix} .2 & .1 \\ 0 & .1732 \end{bmatrix}.$$

The random effects were generated by multiplying $\sigma\mathbf{U}'$ by a column vector of two independent standard normal variables.

A sensible way for fitting nonlinear random effects models is to start by fitting the model with the random effects set to zero. The model is fit by maximum likelihood using a variance function (section 2.3) which is the square of the estimated mean, producing the results in the top part of Table 7.3.

Using the β's estimated without random effects in the model as starting values, two models can then be fit, each with a single random coefficient. These models will contain three nonlinear parameters, β_1, β_2 and u_{11}. If no prior information is available about the variance of the random effect in each model, a natural first guess is $u_{11} = 1$. It is best to avoid first guesses of square roots of variances of zero since when squared, zero may appear to be a minimum to the optimization program when there is a

Table 7.1 *Simulated Michaelis-Menten data with random effects. The rows are data for different subjects.*

Dose								
.0625	.125	.25	.5	1	2	4	8	16
0.032	0.072	0.124	0.286	0.349	0.449	0.541	0.590	0.784
0.100	0.161	0.274	0.420	0.808	1.001	1.043	1.316	1.097
0.071	0.129	0.192	0.405	0.559	0.879	1.010	1.023	1.353
0.064	0.130	0.197	0.409	0.554	0.811	0.967	1.010	1.204
0.069	0.135	0.220	0.316	0.443	0.536	0.732	0.727	0.738
0.057	0.117	0.238	0.297	0.544	0.815	1.070	1.222	1.460
0.065	0.129	0.240	0.403	0.576	0.774	0.932	1.077	1.066
0.063	0.120	0.252	0.355	0.537	0.704	0.682	0.864	0.912
0.059	0.079	0.173	0.257	0.362	0.537	0.487	0.602	0.664
0.045	0.075	0.112	0.249	0.330	0.340	0.436	0.576	0.660
0.074	0.139	0.243	0.373	0.643	0.657	0.772	0.959	0.979
0.053	0.108	0.193	0.312	0.462	0.631	0.734	0.754	0.866
0.056	0.116	0.200	0.329	0.519	0.730	0.843	1.038	0.910
0.068	0.137	0.217	0.428	0.586	0.728	0.997	1.130	1.103
0.044	0.085	0.151	0.191	0.350	0.601	0.532	0.563	0.581
0.067	0.097	0.194	0.331	0.554	0.669	0.964	0.790	1.089
0.053	0.087	0.175	0.251	0.421	0.590	0.664	0.675	0.792
0.070	0.147	0.235	0.306	0.530	0.694	0.698	0.913	0.906
0.043	0.098	0.162	0.340	0.385	0.675	0.633	0.803	0.853
0.057	0.121	0.213	0.323	0.451	0.527	0.605	0.720	0.806
0.042	0.104	0.183	0.298	0.432	0.618	0.741	0.904	0.843
0.053	0.128	0.265	0.341	0.533	0.713	0.841	0.846	0.946
0.038	0.069	0.128	0.300	0.434	0.605	0.680	0.723	0.674
0.070	0.120	0.222	0.347	0.557	0.776	0.902	1.073	0.906
0.069	0.125	0.200	0.381	0.645	0.790	0.851	1.293	1.471
0.050	0.108	0.156	0.257	0.537	0.560	0.674	0.763	0.795
0.066	0.127	0.196	0.354	0.572	0.774	1.016	1.040	1.089
0.058	0.126	0.209	0.359	0.532	0.890	1.036	1.081	1.421
0.055	0.106	0.221	0.334	0.423	0.592	0.654	0.784	0.714
0.046	0.094	0.169	0.316	0.450	0.595	0.620	0.744	0.692

Table 7.2 *Simulated random effects for the Michaelis-Menten data. The rows are data for different subjects.*

γ_1	γ_2	$V = \exp(\gamma_1)$	$K = \exp(\gamma_2)$
-0.200	0.243	0.819	1.275
0.359	-0.022	1.432	0.978
0.303	0.311	1.354	1.365
0.196	0.052	1.217	1.053
-0.244	-0.400	0.784	0.671
0.336	0.276	1.400	1.318
0.190	0.121	1.210	1.128
-0.105	-0.285	0.901	0.752
-0.378	-0.167	0.685	0.846
-0.391	-0.027	0.676	0.973
0.101	-0.068	1.106	0.935
-0.205	-0.163	0.815	0.850
0.103	0.120	1.109	1.128
0.150	0.010	1.161	1.010
-0.298	-0.001	0.742	0.999
0.057	0.043	1.059	1.044
-0.208	-0.004	0.812	0.996
0.040	-0.072	1.041	0.930
-0.157	0.077	0.855	1.080
-0.254	-0.343	0.776	0.710
0.021	0.199	1.021	1.220
0.141	0.088	1.151	1.092
-0.176	0.206	0.839	1.229
0.059	-0.035	1.061	0.966
0.338	0.302	1.402	1.352
-0.127	-0.067	0.881	0.935
0.089	-0.006	1.093	0.994
0.255	0.269	1.290	1.309
-0.249	-0.227	0.779	0.797
-0.123	-0.030	0.884	0.970

Table 7.3 *Results of fitting the Michaelis-Menten data by maximum likelihood.*

	Model 1, no random effects		

$-2 \ln \text{likelihood} = -618.30$

$\hat{\sigma}^2 = 0.0464$ $\text{AIC} = -614.30$

	Lower 95% CL	Estimate	Upper 95% CL
$\hat{\beta}_1$	−.047	.002	.051
$\hat{\beta}_2$	−.074	.009	.091

Model 2, β_1 random

$-2 \ln \text{likelihood} = -866.58$

$\hat{\sigma}^2 = 0.0129$ $\text{AIC} = -860.58$

$\hat{\sigma}\hat{u}_{11} = .187$

	Lower 95% CL	Estimate	Upper 95% CL
$\hat{\beta}_1$	−.097	−.024	.048
$\hat{\beta}_2$	−.060	−.017	.026

Model 3, β_2 random

$-2 \ln \text{likelihood} = -713.67$

$\hat{\sigma}^2 = 0.0256$ $\text{AIC} = -707.67$

$\hat{\sigma}\hat{u}_{11} = .239$

	Lower 95% CL	Estimate	Upper 95% CL
$\hat{\beta}_1$	−.010	.026	.063
$\hat{\beta}_2$	−.037	.069	.176

Model 4, β_1 and β_2 random

$-2 \ln \text{likelihood} = -914.43$

$\hat{\sigma}^2 = 0.00884$ $\text{AIC} = -904.43$

$\hat{\sigma}\hat{u}_{11} = .231$ $\hat{\sigma}\hat{u}_{12} = .117$ $\hat{\sigma}\hat{u}_{22} = .146$

	Lower 95% CL	Estimate	Upper 95% CL
$\hat{\beta}_1$	−.105	−.018	.069
$\hat{\beta}_2$	−.095	−.018	.060

lower minimum at some positive value of u_{11}. If the optimization program produces a negative value of u_{11} as an estimate, this can be replaced by its absolute value since u_{11} is squared when calculating likelihoods.

The results of fitting the two models with one random effect are shown in the middle sections of Table 7.3. The estimated standard deviation of the random effect for a model with a single random effect is $\hat{\sigma}\hat{u}_{11}$. The bottom of Table 7.3 shows the results of fitting the model with two random effects. Model 4 is clearly the best model, as it should be, since this was the model simulated.

The method of estimation is a modification of the method of Lindstrom and Bates (1990). The nonlinear model is expanded with respect to the random effects only. The β's are treated as nonlinear parameters in the optimization. The expansion of the random effects is, first, about their mean values of zero as in the Sheiner and Beal (1980) method. From the linearized equation, estimates of the random effects are obtained for the current values of the β's and **U** matrix. The equation is then linearized again about the estimated values of the random effects as in Lindstrom and Bates' method. This usually gives an improvement in the value of −2 ln likelihood.

The estimation was repeated using REML estimates. The results are shown in Table 7.4. The REML −2 ln likelihoods in Table 7.4 should not be compared with the −2 ln likelihoods in Table 7.3. However, comparing the changes in −2 ln likelihoods in Table 7.3 with the changes of the REML −2 ln likelihoods in Table 7.4, we see that the likelihood ratio test are very similar.

7.2.3 A general nonlinear equation

A general nonlinear equation where the random effects are added to the fixed parameters is

$$y_{ij} = f(x_{ij}, \beta + \gamma_i) + \epsilon_{ij}.$$

The random effects can be added to a subset of the β's by setting the variances of some of the γ's to zero. Take the partial derivatives of the nonlinear function with respect to the γ's that do not have zero variance and expand the nonlinear function in a series

Table 7.4 *Results of fitting the Michaelis-Menten data by REML*

Model 1, no random effects

REML −2 ln likelihood = −603.28

$\hat{\sigma}^2 = 0.0468$ AIC = −599.28

	Lower 95% CL	Estimate	Upper 95% CL
$\hat{\beta}_1$	−.046	.003	.052
$\hat{\beta}_2$	−.072	.011	.094

Model 2, β_1 random

REML −2 ln likelihood = −852.20

$\hat{\sigma}^2 = 0.0130$ AIC = −846.20

$\hat{\sigma}\hat{u}_{11} = .190$

	Lower 95% CL	Estimate	Upper 95% CL
$\hat{\beta}_1$	−.098	−.024	.050
$\hat{\beta}_2$	−.060	−.017	.026

Model 3, β_2 random

REML −2 ln likelihood = −699.55

$\hat{\sigma}^2 = 0.0257$ AIC = −693.55

$\hat{\sigma}\hat{u}_{11} = .244$

	Lower 95% CL	Estimate	Upper 95% CL
$\hat{\beta}_1$	−.010	.026	.063
$\hat{\beta}_2$	−.039	.069	.177

Model 4, β_1 and β_2 random

REML −2 ln likelihood = −901.18

$\hat{\sigma}^2 = 0.00884$ AIC = −891.18

$\hat{\sigma}\hat{u}_{11} = .235$ $\hat{\sigma}\hat{u}_{12} = .120$ $\hat{\sigma}\hat{u}_{22} = .149$

	Lower 95% CL	Estimate	Upper 95% CL
$\hat{\beta}_1$	−.106	−.018	.071
$\hat{\beta}_2$	−.096	−.018	.061

about the current estimates of the fixed and random effects,

$$y_{ij} \sim f(x_{ij}, \hat{\beta} + \hat{\gamma}_i) + \sum_k \frac{\partial f}{\partial \gamma_{ik}} (\gamma_{ik} - \hat{\gamma}_{ik}) + \epsilon_{ij}.$$

Let

$$z_{ijk} = \frac{\partial f}{\partial \gamma_{ik}}.$$

Define the residuals as

$$e_{ij} = y_{ij} - f(x_{ij}, \hat{\beta} + \hat{\gamma}_i) + \sum_k z_{ijk} \hat{\gamma}_{ik}$$

for use in the likelihood, and

$$e_{ij}^{(f)} = y_{ij} - f(x_{ij}, \hat{\beta})$$

for updating $\hat{\gamma}_i$ as in (7.19). Now equations (7.11) to (7.14) or (7.15) can be used to calculate -2 ln likelihood with a three or four iteration inner loop to estimate the random effects using (7.19).

Vonesh and Carter (1992) discuss models that are nonlinear in the fixed parameters, the β's, and linear in the random parameters, the γ_i's. These models can be written

$$y_{ij} = f(x_{ij}, \beta) + \sum_k z_{ijk} \hat{\gamma}_{ik} + \epsilon_{ij}.$$

Since the random effects appear linearly in this model, there is no need to expand the function in a series about the current values of the random effects. These models are much easier to handle than models where the random effects appear in a truly nonlinear fashion. The β's can be included as extra nonlinear parameters in the nonlinear optimization. At any stage of the search for the maximum likelihood estimates of the parameters, let $\hat{\boldsymbol{\beta}}$ denote the current values of the fixed parameters. Form the residuals

$$e_{ij}^{(f)} = y_{ij} - f(x_{ij}, \hat{\beta}).$$

Writing these residuals as a vector for each subject, the model can be written as a linear Laird-Ware model without fixed effects,

$$\mathbf{e}_i^{(f)} = \mathbf{Z}_i \boldsymbol{\gamma}_i + \boldsymbol{\epsilon}_i.$$

The nonlinear parameters in this model are the β's and the elements of $\mathbf{U}$, the Cholesky factor of the covariance matrix of $\boldsymbol{\gamma}_i$,

$$\text{cov}(\boldsymbol{\gamma}_i) = \sigma^2 \mathbf{B} = \sigma^2 \mathbf{U}'\mathbf{U}.$$

7.3 A four parameter logistic model

If equation (7.2) is plotted as a function of $\ln x$, the result is a logistic shaped curve (Rudemo, Ruppert and Streibig, 1989). Let $u_j = \ln x_j$, then

$$y_j = \frac{\exp(\beta_1)}{1 + \exp(\beta_2 - u_j)} + \epsilon_j.$$

Another nonlinear model commonly used in assays is a four parameter logistic dose response curve. One parameterization of this model is

$$y_j = \beta_1 + \beta_4 \frac{[(x_j/\exp(\beta_3))]^{\beta_2}}{1 + [(x_j/\exp(\beta_3))]^{\beta_2}} + \epsilon_j,$$

where $\beta_1 \geq 0$ is the lower asymptote, $\beta_4 > 0$ is the distance between the asymptotes, $\beta_2 > 0$ is the slope parameter, β_3 is the natural log of the value of the dose, x, where the curve is half way between the lower asymptote and the upper asymptote. This is known as the 50% effective dose and is referred to as the ED50. This model is nonlinear in the two parameters, β_2 and β_3. A graph is shown in Figure 7.2.

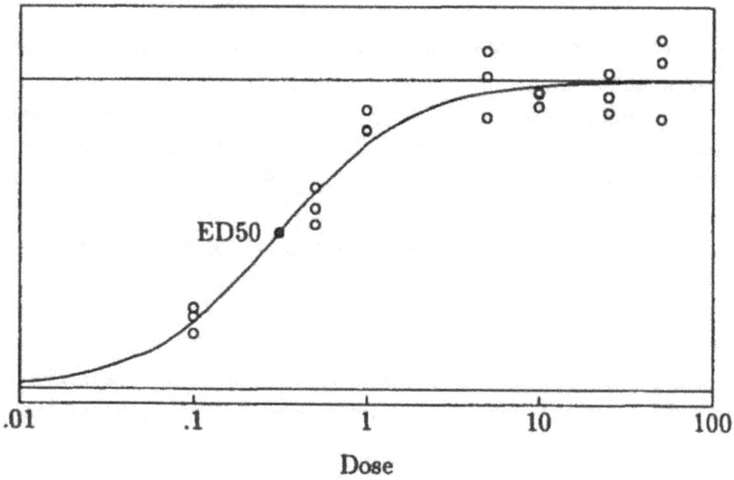

Fig. 7.2 Dose response curve

The lower asymptote, β_1, will usually not contain a random effect since it is determined by some background level. However, the upper asymptote, β_4, may well have a random component. Different 'subjects' or cell cultures may asymptote at a different level,

$$y_j = \beta_1 + (\beta_4 + \gamma_4)\frac{[(x_j/\exp(\beta_3)]^{\beta_2}}{1 + [(x_j/\exp(\beta_3)]^{\beta_2}} + \epsilon_j.$$

If β_4 is the only parameter with a random component, the problem is not difficult since the model is only nonlinear with respect to the parameters β_2 and β_3. If β_2 and β_3 are treated as nonlinear parameters by the optimization program (as opposed to linearizing with respect to β_2 and β_3), for given values of β_2 and β_3 at any stage of the optimization, the problem is linear with respect to the remaining parameters, including the random effect on β_4.

The other parameter that often has a random component is β_3, the ln ED50,

$$y_j = \beta_1 + (\beta_4 + \gamma_4)\frac{[(x_j/\exp(\beta_3 + \gamma_3)]^{\beta_2}}{1 + [(x_j/\exp(\beta_3 + \gamma_3)]^{\beta_2}} + \epsilon_j.$$

This is the truly nonlinear case where it is necessary to expand the function in a series about γ_3.

CHAPTER 8

Multivariate Models

Jones (1984) and Jones and Tryon (1987) considered a class of state space models for handling multivariate observations at unequally spaced times. Also included is the situation where some of the multivariate observations are missing. For example, if three variables are measured at each observation time, sometimes one or two of the three variables may be unobserved. The most general model considered in the paper, in state space form, was

$$d\begin{bmatrix} \mathbf{s}(t) \\ \mu \end{bmatrix} = \begin{bmatrix} \mathbf{A} & 0 \\ 0 & 0 \end{bmatrix} \begin{bmatrix} \mathbf{s}(t) \\ \mu \end{bmatrix} dt + \begin{bmatrix} \mathbf{G}d\mathbf{w}(t) \\ 0 \end{bmatrix}, \qquad (8.1)$$

where $\mathbf{s}(t)$ is a $d \times 1$ vector denoting the state of the system expressed as deviations from a constant mean vector, μ. The random input to the state, $d\mathbf{w}(t)$, is the differential of a multivariate Wiener process, $\mathbf{w}(t)$, which produces continuous time multivariate 'white noise'. The dimension of $\mathbf{w}(t)$ is $p \times 1$ where $p \leq d$. The covariance matrix of the random input is

$$E|d\mathbf{w}(t)d\mathbf{w}'(t)| = \mathbf{I}dt.$$

Each component has unit variance per unit time, and the components are uncorrelated. To introduce correlation into the random input to the state, the random input is premultipled by a $d \times p$ matrix $\mathbf{G}$, the components of which are to be estimated. $\mathbf{A}$ is a $d \times d$ matrix of constants which also are to be estimated. $\mathbf{s}(t)$ will be a multivariate stationary autoregression of order one, MAR(1), if the roots of

$$|\mathbf{I} - \mathbf{A}z| = 0$$

have negative real parts which implies that the eigenvalues of $\mathbf{A}$ have negative real parts. If $\mathbf{A} = 0$, $\mathbf{s}(t)$ will be a correlated multivariate random walk.

If there are unequally spaced observations at times t_1, t_2, $\cdots$, t_n, the observation equation for one of these times is

$$\mathbf{y}(t_j) = \mathbf{s}(t_j) + \boldsymbol{\mu} + \mathbf{v}(t_j), \tag{8.2}$$

where $\mathbf{v}(t_j)$ is multivariate observational error, uncorrelated with $\mathbf{w}(t)$, and with mean zero and covariance matrix $\mathbf{R}$. The state space model defined by equations (8.1) and (8.2) provides the seed for generalizing the Laird-Ware model (2.1) to multivariate observations at unequally spaced observation times and with some of the components missing. Various types of within subject error structure are also possible.

A multivariate generalization of the Laird-Ware model can be written the same as equation (2.1), where the data have a block structure. Suppose that at each observation time there is the potential of observing d (for dimension) different variables on a subject. The word 'potential' is used since perhaps all d variables are not observed at each observation time. Typically, d will be small such as two or three. Also, the observation times may be different for each subject, or the different variables on a subject may not all be observed at the same times.

There are several possible structures for the within subject errors, $\boldsymbol{\epsilon}_i$. The equivalent structure to independent within subject errors in the univariate case is a block diagonal structure for $\mathbf{W}_i$. If all the variables are observed at an observation time, that block of $\mathbf{W}_i$ would have an arbitrary $d \times d$ covariance matrix. If only a subset of the variables are observed, $\mathbf{W}_i$ would be the corresponding reduced matrix. Another possible structure is a multivariate AR(1) structure as was used by Jones (1984) in the single subject case. A third possibility is a multivariate random walk or integrated random walk.

Consider, first, a multivariate, single group, repeated measures experiment. For each subject, under each treatment condition, two or three response variables are measured. The model for subject i under treatment condition j is

$$\mathbf{y}_{ij} = \boldsymbol{\beta}_j + \boldsymbol{\gamma}_i + \boldsymbol{\epsilon}_{ij}.$$

All the elements of this equation are column vectors of length d. $\mathbf{y}_{ij}$ is the vector of responses for subject i under treatment condition j, $\boldsymbol{\beta}_j$ is a vector of fixed means for the responses to treatment condition j, $\boldsymbol{\gamma}_i$ is a vector of the random subject effect for subject i with a general between subject covariance matrix $\sigma^2\mathbf{B}$, and ϵ_{ij} is a vector of random within subject errors with a general covariance matrix $\sigma^2\mathbf{W}$. If there are missing observations in the response vector, the model can be written

$$\mathbf{y}_{ij} = \mathbf{X}_{ij}\boldsymbol{\beta}_j + \mathbf{X}_{ij}\boldsymbol{\gamma}_i + \mathbf{X}_{ij}\boldsymbol{\epsilon}_{ij},$$

where $\mathbf{X}_{ij} = \mathbf{I}$, the identity matrix if there are no missing observations, and, if there are missing observations, $\mathbf{X}_{ij}$ is the identity matrix with rows removed corresponding to the missing observations. For example, if there are three response variables and the second one is missing,

$$\mathbf{X}_{ij} = \begin{bmatrix} 1 & 0 & 0 \\ 0 & 0 & 1 \end{bmatrix}.$$

If all the observations for one subject are concatenated into a single vector, the model for subject i is

$$\begin{bmatrix} \mathbf{y}_{i1} \\ \vdots \\ \mathbf{y}_{in} \end{bmatrix} = \begin{bmatrix} \mathbf{X}_{i1} & & 0 \\ & \ddots & \\ 0 & & \mathbf{X}_{in} \end{bmatrix} \begin{bmatrix} \boldsymbol{\beta}_1 \\ \vdots \\ \boldsymbol{\beta}_n \end{bmatrix} + \begin{bmatrix} \mathbf{X}_{i1} \\ \vdots \\ \mathbf{X}_{in} \end{bmatrix} \boldsymbol{\gamma}_i + \begin{bmatrix} \mathbf{X}_{i1}\boldsymbol{\epsilon}_{i1} \\ \vdots \\ \mathbf{X}_{in}\boldsymbol{\epsilon}_{in} \end{bmatrix},$$

where n is the number of treatment conditions.

If there are no missing observations in this balanced design, the total error covariance matrix for subject i involving both random vectors is

$$\mathbf{V}_i = \sigma^2 \begin{bmatrix} \mathbf{W}+\mathbf{B} & \mathbf{B} & \cdots & \mathbf{B} \\ \mathbf{B} & \mathbf{W}+\mathbf{B} & \cdots & \mathbf{B} \\ \vdots & \vdots & & \vdots \\ \mathbf{B} & \mathbf{B} & \cdots & \mathbf{W}+\mathbf{B} \end{bmatrix}.$$

For the case of missing observations, the matrices will be reduced by the design matrices $\mathbf{X}_{ij}$. Let

$$\mathbf{B}_{ijk} = \mathbf{X}_{ij}\mathbf{B}\mathbf{X}'_{ik} \qquad \text{and} \qquad \mathbf{W}_{ij} = \mathbf{X}_{ij}\mathbf{W}\mathbf{X}'_{ij},$$

then the error covariance matrix for subject i becomes

$$\mathbf{V}_i = \sigma^2 \begin{bmatrix} \mathbf{W}_{i1}+\mathbf{B}_{i11} & \mathbf{B}_{i12} & \cdots & \mathbf{B}_{i1n} \\ \mathbf{B}_{i21} & \mathbf{W}_{i2}+\mathbf{B}_{i22} & \cdots & \mathbf{B}_{i2n} \\ \vdots & \vdots & & \vdots \\ \mathbf{B}_{in1} & \mathbf{B}_{in2} & \cdots & \mathbf{W}_{in}+\mathbf{B}_{inn} \end{bmatrix}.$$

Assuming that the errors have Gaussian distributions, −2 ln likelihood can be calculated as a function of the regression coefficients, σ^2 and the matrices **B** and **W**. For given values of **B** and **W**, the regression coefficients can be estimated by weighted least squares, and σ^2 estimated from the mean square error. Substituting these estimates back into −2 ln likelihood gives −2 ln likelihood concentrated with respect to the regression coefficients as a function of **B** and **W**. Expressing the covariance matrices in factored form,

$$\mathbf{B} = \mathbf{U}_b'\mathbf{U}_b \qquad \text{and} \qquad \mathbf{W} = \mathbf{U}_w'\mathbf{U}_w,$$

where $\mathbf{U}_b$ and $\mathbf{U}_w$ are $d \times d$ upper triangular matrices, nonlinear optimization (Dennis and Schnabel, 1983) can be used to obtain maximum likelihood estimates of the elements of the **U** matrices. In the bivariate case, $d = 2$, there are 6 elements that need to be estimated using nonlinear optimization. If $d = 3$, there are 12 elements. Optimizing with respect to the elements of the **U** matrices ensures that the matrices **B** and **W** will remain non-negative definite and therefore are in the parameter space.

The hypothesis of no treatment effect is

$$H_0: \boldsymbol{\beta}_1 = \boldsymbol{\beta}_2 = \cdots = \boldsymbol{\beta}_n,$$

which can be tested as a contrast or by fitting the reduced model

$$\begin{bmatrix} \mathbf{y}_{i1} \\ \vdots \\ \mathbf{y}_{in} \end{bmatrix} = \begin{bmatrix} \mathbf{X}_{i1} \\ \vdots \\ \mathbf{X}_{in} \end{bmatrix} \boldsymbol{\beta} + \begin{bmatrix} \mathbf{X}_{i1} \\ \vdots \\ \mathbf{X}_{in} \end{bmatrix} \boldsymbol{\gamma}_i + \begin{bmatrix} \mathbf{X}_{i1}\boldsymbol{\epsilon}_{i1} \\ \vdots \\ \mathbf{X}_{in}\boldsymbol{\epsilon}_{in} \end{bmatrix},$$

where $\boldsymbol{\beta}$ is a $d \times 1$ vector of the means of the response variables.

This multivariate extension of a standard single group repeated measures experiment with missing observation is fairly straight forward; however, it sets the stage for more general designs. Covariates can be included. If there are two groups of subjects, introduce an indicator covariate, c, for each subject, taking the value 0 for one group and 1 for the other. The model including a group effect is

$$\begin{bmatrix} \mathbf{y}_{i1} \\ \vdots \\ \mathbf{y}_{in} \end{bmatrix} = \begin{bmatrix} \mathbf{X}_{i1} & & 0 & c\mathbf{X}_{i1} \\ & \ddots & \vdots & \vdots \\ 0 & & \mathbf{X}_{in} & c\mathbf{X}_{in} \end{bmatrix} \begin{bmatrix} \boldsymbol{\beta}_1 \\ \vdots \\ \boldsymbol{\beta}_n \\ \boldsymbol{\beta}_g \end{bmatrix}$$

$$+ \begin{bmatrix} \mathbf{X}_{i1} \\ \vdots \\ \mathbf{X}_{in} \end{bmatrix} \gamma_i + \begin{bmatrix} \mathbf{X}_{i1}\epsilon_{i1} \\ \vdots \\ \mathbf{X}_{in}\epsilon_{in} \end{bmatrix},$$

where $\boldsymbol{\beta}_g$ is a $d \times 1$ vector of the vertical shift between groups for the d response variables. The complete model with both group and group by treatment interaction is

$$\begin{bmatrix} \mathbf{y}_{i1} \\ \vdots \\ \mathbf{y}_{in} \end{bmatrix} = \begin{bmatrix} \mathbf{X}_{i1} & & 0 & c\mathbf{X}_{i1} & & 0 \\ & \ddots & \vdots & & \ddots & \vdots \\ 0 & & \mathbf{X}_{in} & 0 & & c\mathbf{X}_{in} \end{bmatrix} \begin{bmatrix} \boldsymbol{\beta}_1 \\ \vdots \\ \boldsymbol{\beta}_n \\ \boldsymbol{\beta}_{n+1} \\ \vdots \\ \boldsymbol{\beta}_{2n} \end{bmatrix}$$

$$+ \begin{bmatrix} \mathbf{X}_{i1} \\ \vdots \\ \mathbf{X}_{in} \end{bmatrix} \gamma_i + \begin{bmatrix} \mathbf{X}_{i1}\epsilon_{i1} \\ \vdots \\ \mathbf{X}_{in}\epsilon_{in} \end{bmatrix}.$$

Here, the lower half of the $\boldsymbol{\beta}$ vector is the difference in the regression coefficients for the two groups.

A multivariate growth curve model with random effects is easily demonstrated by considering the special case of bivariate responses and straight lines in an unbalanced situation. If the observations for subject i are taken at times $t_{i1}, t_{i2}, \cdots, t_{in_i}$, these times may be different for different subjects. The bivariate model, where there is a different straight line for each of the two response variables and a different random subject effect for each response variable consisting of a random vertical shift, is

$$\begin{bmatrix} y_{i1}(t_{i1}) \\ y_{i2}(t_{i1}) \\ y_{i1}(t_{i2}) \\ y_{i2}(t_{i2}) \\ y_{i1}(t_{i3}) \\ y_{i2}(t_{i3}) \\ \vdots \end{bmatrix} = \begin{bmatrix} 1 & t_{i1} & 0 & 0 \\ 0 & 0 & 1 & t_{i1} \\ 1 & t_{i2} & 0 & 0 \\ 0 & 0 & 1 & t_{i2} \\ 1 & t_{i3} & 0 & 0 \\ 0 & 0 & 1 & t_{i3} \\ & \vdots & & \end{bmatrix} \begin{bmatrix} \beta_1 \\ \beta_2 \\ \beta_3 \\ \beta_4 \end{bmatrix}$$

$$
+\begin{bmatrix} 1 & 0 \\ 0 & 1 \\ 1 & 0 \\ 0 & 1 \\ 1 & 0 \\ 0 & 1 \\ \vdots & \end{bmatrix} \begin{bmatrix} \gamma_{i1} \\ \gamma_{i2} \end{bmatrix} + \begin{bmatrix} \epsilon_{i1}(t_{i1}) \\ \epsilon_{i2}(t_{i1}) \\ \epsilon_{i1}(t_{i2}) \\ \epsilon_{i2}(t_{i2}) \\ \epsilon_{i1}(t_{i3}) \\ \epsilon_{i2}(t_{i3}) \\ \vdots \end{bmatrix}.
$$

If some of the observations are missing, the corresponding row of this equation is deleted.

The above two examples demonstrate that it is not difficult to extend the Laird-Ware model to multivariate situations, even unbalanced situations, by considering the model to have a block structure. The difficulty is the modelling of within subject serial correlation in the multivariate case. If serial correlation is present and not modelled, inferences from the analysis will not be correct. The use of multivariate autoregressions is one way to model serial correlation.

8.1 Continuous time multivariate AR(1) errors

For equally spaced observations, without missing data, the use of multivariate autoregressions has a long history (Whittle, 1963; Jones, 1964). A first order, discrete time, bivariate, autoregression for the within subject errors is

$$
\begin{bmatrix} \epsilon_{i1}(j+1) \\ \epsilon_{i2}(j+1) \end{bmatrix} = \begin{bmatrix} \phi_{11} & \phi_{12} \\ \phi_{21} & \phi_{22} \end{bmatrix} \begin{bmatrix} \epsilon_{i1}(j) \\ \epsilon_{i2}(j) \end{bmatrix} + \begin{bmatrix} \eta_{i1}(j+1) \\ \eta_{i2}(j+1) \end{bmatrix},
$$

where the vectors of random inputs, $\eta_i(j+1)$ are uncorrelated at different times and have an arbitrary covariance matrix that is constant over time. The process is stationary or stable if the roots of $|\mathbf{I} - \mathbf{\Phi} z| = 0$ are outside the unit circle which implies that the eigenvalues of $\mathbf{\Phi}$ are inside the unit circle.

A multivariate *continuous time* autoregression must be used when there are unequally spaced observations. The multivariate continuous time first order autoregression, MCAR(1), with zero mean is characterized by the system of stochastic first order differential equations, as in equation (8.1),

$$
d\mathbf{s}(t) = \mathbf{A}\mathbf{s}(t)dt + \mathbf{G}d\mathbf{w}(t). \qquad (8.3)
$$

This process will be stationary or stable if the roots of $|\mathbf{I} - \mathbf{A}z|$ have negative real parts which implies that the eigenvalues of $\mathbf{A}$ have negative real parts. If the time interval between two observations is δt, the discrete time autoregressive matrix for this time interval can be calculated from $\mathbf{A}$ as

$$\mathbf{\Phi}(\delta t) = e^{\mathbf{A}\delta t},$$

(Jones, 1984). The calculation of $\mathbf{\Phi}$ requires the calculation of a matrix exponential (Moler and Van Loan, 1978). A possible method of calculating a matrix exponential is to perform an eigenvalue, eigenvector decomposition of the matrix $\mathbf{A}$,

$$\mathbf{A} = \mathbf{C}\mathbf{\Lambda}\mathbf{C}^{-1}, \tag{8.4}$$

where $\mathbf{\Lambda}$ is a diagonal matrix of the eigenvalues of $\mathbf{A}$, and $\mathbf{C}$ is a $d \times d$ matrix , the columns of which are right eigenvectors of $\mathbf{A}$. The eigenvalues and eigenvectors may be complex. The discrete time autoregressive matrix for the time step of δt is

$$\mathbf{\Phi}(\delta t) = \mathbf{C}e^{\mathbf{\Lambda}\delta t}\mathbf{C}^{-1}. \tag{8.5}$$

In this equation $e^{\mathbf{\Lambda}\delta t}$ is a diagonal matrix of e raised to the powers $\lambda_k \delta t$, where the λ_k are the eigenvalues.

Having integrated over a time step δt, we can now write the continuous time state equation (8.3) as a discrete time state equation for the given time step,

$$\mathbf{s}(t) = \mathbf{\Phi}(\delta t)\mathbf{s}(t - \delta t) + \boldsymbol{\eta}(\delta t), \tag{8.6}$$

where $\boldsymbol{\eta}(\delta t)$ is the random input in the interval δt with covariance matrix $\mathbf{Q}(\delta t)$. Substituting (8.5) into (8.6) gives

$$\mathbf{s}(t) = \mathbf{C}e^{\mathbf{\Lambda}\delta t}\mathbf{C}^{-1}\mathbf{s}(t - \delta t) + \boldsymbol{\eta}(\delta t). \tag{8.7}$$

Premultplying (8.7) by $\mathbf{C}^{-1}$ gives

$$\mathbf{C}^{-1}\mathbf{s}(t) = e^{\mathbf{\Lambda}\delta t}\mathbf{C}^{-1}\mathbf{s}(t - \delta t) + \mathbf{C}^{-1}\boldsymbol{\eta}(\delta t).$$

Now define a rotated state vector which may be complex,

$$\mathbf{s}_r(t) = \mathbf{C}^{-1}\mathbf{s}(t).$$

The rotated state vector can be calculated by solving a system of complex equations,

$$\mathbf{C}\mathbf{s}_r(t) = \mathbf{s}(t).$$

The state equation for the rotated state vector is

$$\mathbf{s}_r(t) = e^{\mathbf{A}\delta t}\mathbf{s}_r(t - \delta t) + \boldsymbol{\eta}_r(\delta t), \tag{8.8}$$

where $\boldsymbol{\eta}_r(\delta t) = \mathbf{C}^{-1}\boldsymbol{\eta}(\delta t)$ is the random error introduced into the rotated state equation and may be complex. The state transition matrix in this equation is diagonal and possibly complex. Since the state transition equation is diagonal, the rotated state vector is said to be uncoupled since each element is predicted independently of the other elements. The covariance matrix of the rotated state vector, (8.8), is

$$\mathbf{Q}_r(\delta t) = \mathbf{C}^{-1}\mathbf{Q}(\delta t)(\mathbf{C}^*)^{-1},$$

where * denotes the complex conjugate transposed matrix, and the inverse transformation is

$$\mathbf{Q}(\delta t) = \mathbf{C}\mathbf{Q}_r(\delta t)\mathbf{C}^*. \tag{8.9}$$

The covariance matrix of the random input to the state equation depends on the length of the time intervals between observations. The random input over an interval of length δt is (Kalman and Bucy, 1961)

$$\int_0^{\delta t} \boldsymbol{\Phi}(\delta t - t)\mathbf{G}d\mathbf{w}(t),$$

and has covariance matrix

$$\mathbf{Q}(\delta t) = \int_0^{\delta t} \boldsymbol{\Phi}(\delta t - t)\mathbf{G}\mathbf{G}'\boldsymbol{\Phi}'(\delta t - t)dt. \tag{8.10}$$

The calculation of $\boldsymbol{\Phi}(\delta t)$ and $\mathbf{Q}(\delta t)$ for any step size, δt, allows the discrete time Kalman filter (Kalman, 1960) to be used for calculating the exact likelihood (Schweppe, 1965) for arbitrary spacing of the observations.

When using the Kalman filter, the calculations can be carried out using the rotated state and its rotated covariance matrix using complex arithmetic. In the applications presented in this book, the state corresponds to the within subject errors augmented by the between subject random effects. In this section I will just consider the within subject errors. In these applications, the initial state vector for each subject is a vector of zeros since the within subject errors have mean zero. The rotated initial state vector is also a vector of zeros.

For a stationary process, the initial state covariance matrix is the limit of $\mathbf{Q}(\delta t)$ as $\delta t \to \infty$,

$$\mathbf{P}(0|0) = \mathbf{Q}(\infty).$$

$\mathbf{Q}(\delta t)$ can be evaluated using the eigenvalues and eigenvectors of $\mathbf{A}$ as in Jones (1984). Substituting (8.5) into (8.10) gives

$$\mathbf{Q}(\delta t) = \int_0^{\delta t} \mathbf{C} e^{\mathbf{A}(\delta t - t)} \mathbf{C}^{-1} \mathbf{G}\mathbf{G}'(\mathbf{C}^*)^{-1} e^{\mathbf{A}^*(\delta t - t)} \mathbf{C}^* dt, \quad (8.11)$$

and, the covariance matrix for the random input to the rotated state equation is

$$\mathbf{Q}_r(\delta t) = \int_0^{\delta t} e^{\mathbf{A}(\delta t - t)} \mathbf{C}^{-1} \mathbf{G}\mathbf{G}'(\mathbf{C}^*)^{-1} e^{\mathbf{A}^*(\delta t - t)} dt, \quad (8.12)$$

This rotated covariance matrix may be complex Hermitian, that is $\mathbf{Q}_r(\delta t) = \mathbf{Q}_r^*(\delta t)$, the matrix equals its complex conjugate transposed matrix. Let

$$\mathbf{K} = \mathbf{C}^{-1}\mathbf{G}\mathbf{G}'(\mathbf{C}^*)^{-1}, \quad (8.13)$$

a complex matrix that can be calculate from $\mathbf{G}$ and the matrix of eigenvectors. Now (8.12) can be integrated element by element in terms of the elements of K giving for the elements of $\mathbf{Q}_r(\delta t)$,

$$Q_{jk}^{(r)}(\delta t) = K_{jk} \frac{e^{(\lambda_j + \bar{\lambda}_k)\delta t} - 1}{\lambda_j + \bar{\lambda}_k}, \quad \lambda_j + \bar{\lambda}_k \neq 0,$$

$$Q_{jk}^{(r)}(\delta t) = K_{jk}\delta t, \quad \lambda_j + \bar{\lambda}_k = 0, \quad (8.14)$$

where the $\bar{\lambda}_k$ denotes complex conjugate. If the process is stationary, the eigenvalues, λ_j have negative real parts, and the limit of $Q_{jk}^{(r)}(\delta t)$ as $\delta t \to \infty$ is

$$\mathbf{P}_r(0|0) = Q_{jk}^{(r)}(\infty) = -\frac{K_{jk}}{\lambda_j + \bar{\lambda}_k}. \quad (8.15)$$

This gives the initial state covariance matrix of the rotated state vector for each subject.

If $\lambda_j + \bar{\lambda}_k$ has a non-negative real part, the process is non-stationary, and the jkth element of $\mathbf{Q}_r(\infty)$ is ∞ representing a diffuse prior. One possibility is to set this element to a 'large' number. Because the components of a multivariate response vector may be on different scales, e.g. one may be on the order of 10,000 and another may be on an order of 0.001, 'large' may be different for the two components. If an element of the $\mathbf{Q}_r(\infty)$ matrix is set too large, all significant digits may be lost in the subtraction of step 8 of the Kalman recursion. This is the step where the state covariance matrix is updated, and the calculation involves a subtraction. A possible way to define a large number for each component is to calculate the sample standard deviation of each component based on the raw data. Let this estimated standard deviation for component j be sd_j. A 'large' value for the jkth element of $\mathbf{Q}(\infty)$ could be $(sd_j)(sd_k)10^4$ if the calculations are carried out in double precision with 15 significant digits. In order to get an element of $\mathbf{Q}_r(\infty)$, when $\lambda_j + \bar{\lambda}_k$ has a non-negative real part, construct a column vector $\mathbf{l}$ with elements $10^2(sd_j)$, and calculate $\mathbf{l}_r = \mathbf{C}^{-1}\mathbf{l}$. The corresponding element of $\mathbf{Q}_r(\infty)$ is then $l_j l_k$.

When using the Kalman filter on the rotated state vector, the complex version of the Kalman recursion, with the constant vectors $\mathbf{f}_s$ and $\mathbf{f}_o$ removed, becomes:

1. Calculate a one step prediction using complex arithmetic,

$$\mathbf{s}_r(t|t-\delta t) = e^{\mathbf{A}\delta t}\mathbf{s}_r(t-\delta t|t-\delta t).$$

2. Calculate the complex (Hermitian) covariance matrix of this prediction,

$$\mathbf{P}_r(t|t-\delta t) = e^{\mathbf{A}\delta t}\mathbf{P}_r(t-\delta t|t-\delta t)e^{\mathbf{A}^*\delta t} + \mathbf{Q}_r(\delta t).$$

3. Calculate the prediction of the next observation vector. It is necessary to rotate the state vector back to the original coordinate system to predict the next observation vector. Suppose first that the observations are all the elements of the state vector. In this case the equation is

$$\mathbf{y}(t|t-\delta t) = \mathbf{C}\mathbf{s}_r(t|t-\delta t).$$

If some of the observations are missing, the corresponding rows of the $\mathbf{C}$ matrix are deleted. In the more general case where a linear combination of the elements of the state vector are observed corresponding to the matrix $\mathbf{H}(t)$, the equation is

$$\mathbf{y}(t|t-\delta t) = \mathbf{H}(t)\mathbf{C}\mathbf{s}_r(t|t-\delta t).$$

4. Calculate the innovation vector as before,

$$\mathbf{I}(t) = \mathbf{y}(t) - \mathbf{y}(t|t-\delta t).$$

5. Calculate the covariance matrix of the innovation vector which is real, although the calculations on the right hand side are complex,

$$\mathbf{V}(t) = \mathbf{H}(t)\mathbf{C}\mathbf{P}_r(t|t-\delta t)\mathbf{C}^*\mathbf{H}'(t) + \mathbf{R}(t).$$

6. Accumulate the quantities needed to calculate −2 ln likelihood at the end of the recursion,

$$RSS \leftarrow RSS + \mathbf{I}'(t)\mathbf{V}^{-1}(t)\mathbf{I}(t),$$

$$\Delta \leftarrow \Delta + \ln|\mathbf{V}(t)|$$

This step is not changed.

7. To update the estimate of the state vector, let

$$\mathbf{A}(t) = \mathbf{H}(t)\mathbf{C}\mathbf{P}_r(t|t-\delta t),$$

a complex matrix. Then

$$\mathbf{s}_r(t|t) = \mathbf{s}_r(t|t-\delta t) + \mathbf{A}^*(t)\mathbf{V}^{-1}(t)\mathbf{I}(t),$$

8. The updated covariance matrix of the rotated state is,

$$\mathbf{P}_r(t|t) = \mathbf{P}_r(t|t-\delta t) - \mathbf{A}^*(t)\mathbf{V}^{-1}(t)\mathbf{A}(t).$$

8.2 Multivariate mixed models

The use of a state space representation allows the modelling of various within subject error structures. Assuming Gaussian errors, the Kalman filter can be used to calculate the exact likelihood for given values of the nonlinear parameters. A nonlinear optimization program can be used to obtain maximum likelihood estimates of the parameters.

A multivariate version of equation (2.1) can be written

$$\mathbf{y}(t_{ij}) = \mathbf{X}_i(t_{ij})\beta + \mathbf{Z}_i(t_{ij})\gamma_i + \epsilon(t_{ij}), \tag{8.16}$$

where $\mathbf{y}(t_{ij})$ is the portion of the $\mathbf{y}$ vector for subject i corresponding to the multivariate responses at time t_{ij}. $\mathbf{X}_i(t_{ij})$, $\mathbf{Z}_i(t_{ij})$ and $\epsilon(t_{ij})$ are the corresponding rows of the $\mathbf{X}_i$, $\mathbf{Z}_i$ and ϵ vectors. Suppose that $\epsilon(t_{ij})$ consists of an underlying continuous time multivariate first order autoregression, MCAR(1), process plus additive random noise. The model can be written in state space form. The continuous time state equation for subject i is,

$$d\begin{bmatrix} \epsilon_i(t) \\ \gamma_i \end{bmatrix} = \begin{bmatrix} \mathbf{A} & 0 \\ 0 & 0 \end{bmatrix} \begin{bmatrix} \epsilon_i(t) \\ \gamma_i \end{bmatrix} dt + \begin{bmatrix} \mathbf{G}d\mathbf{w}_i(t) \\ 0 \end{bmatrix}, \tag{8.17}$$

and the observation equation for an observation at time t_{ij} is

$$\xi_i(t_{ij}) = [\mathbf{I} \quad \mathbf{Z}_i(t_{ij})] \begin{bmatrix} \epsilon_i(t_{ij}) \\ \gamma_i \end{bmatrix} + \mathbf{v}_i(t_{ij}).$$

In these equations, $\epsilon_i(t)$ is the underlying MCAR(1) process, and $\mathbf{v}_i(t_{ij})$ is an additive random vector.

The power of this multivariate model is the possibility of determining cause and effect relationships among the response variables. Also, missing observations in the response variables cause no problems. A non-zero off diagonal element in the $\mathbf{A}$ matrix indicates that the rate of change of one variable is affected by the level of another variable. An off diagonal element in the $\mathbf{G}$ matrix may indicate that a variable not in the model is affecting two of the response variables.

8.3 The multivariate algorithm

The algorithm given in Section 5.1.1 needs some modification for the multivariate case. At each observation time, a matrix should be set up in storage which is the current rows of the $\mathbf{X}$ matrix of the regression augmented by the $\mathbf{y}$ vector,

$$[\ \mathbf{x}(t_j) \quad \mathbf{y}(t_j)\ ].$$

In the univariate case, this was a row vector. Now, the number of rows is the dimension of the multivariate model, d. Initialize a scalar Δ and a $(b+1) \times (b+1)$ matrix $\mathbf{M}$ to zero for accumulating the determinant and sums of products for the regression. The state and state transition matrix are

$$\mathbf{S}(t) = \begin{bmatrix} \mathbf{s}_r(t) \\ \gamma_i \end{bmatrix} \qquad \mathbf{\Phi}(\delta t) = \begin{bmatrix} e^{\mathbf{A}\delta t} & 0 \\ 0 & \mathbf{I} \end{bmatrix},$$

and the initial state covariance matrix and the covariance matrix of the random input to the state for a time step of δt are

$$\mathbf{P}(0|0) = \begin{bmatrix} \mathbf{P}_r(0|0) & 0 \\ 0 & \sigma^2\mathbf{B} \end{bmatrix} \qquad \mathbf{Q}(\delta t) = \begin{bmatrix} \mathbf{Q}_r(\delta t) & 0 \\ 0 & 0 \end{bmatrix}.$$

The steps of the modified Kalman recursion which concentrates regression coefficients out of the likelihood for multivariate data are

1. Calculate a one step prediction,

$$\mathbf{S}(t_j|t_j - \delta t) = \mathbf{\Phi}(\delta t)\mathbf{S}(t_j - \delta t|t_j - \delta t).$$

2. Calculate the covariance matrix of this prediction,

$$\mathbf{P}(t_j|t_j - \delta t) = \mathbf{\Phi}(\delta t)\mathbf{P}(t_j - \delta t|t_j - \delta t)\mathbf{\Phi}'(\delta t) + \mathbf{Q}(\delta t).$$

3. Calculate the prediction of the next observation matrix,

$$\begin{bmatrix} \mathbf{x}(t_j|t_j - \delta t) & \mathbf{y}(t_j|t_j - \delta t) \end{bmatrix} = \mathbf{H}(t_j)\mathbf{S}(t_j|t_j - \delta t),$$

where $\mathbf{H}(t_j) = [\mathbf{C} \quad \mathbf{Z}_i(t_{ij})]$, and $\mathbf{C}$ is the matrix of right eigenvectors of $\mathbf{A}$. If there are missing observations in the $\mathbf{y}(t_j)$ vector, or in the $\mathbf{x}(t_j)$ matrix, the corresponding row or rows of $\mathbf{H}(t_j)$ are removed. A row of $b+1$ predictions is produced for each dependent variable.

4. Calculate the innovation *matrix* which is the difference between the rows of the $\mathbf{X}$ augmented by $\mathbf{y}$ matrix and the prediction of these rows,

$$\mathbf{I}(t_j) = \begin{bmatrix} \mathbf{x}(t_j) & \mathbf{y}(t_j) \end{bmatrix} - \begin{bmatrix} \mathbf{x}(t_j|t_j - \delta t) & \mathbf{y}(t_j|t_j - \delta t) \end{bmatrix}.$$

5. Calculate the covariance matrix of the innovation vector,

$$\mathbf{V}(t_j) = \mathbf{H}(t_j)\mathbf{P}(t_j|t_j - \delta t)\mathbf{H}'(t_j) + \mathbf{R}(t_j).$$

   The observational error variance, $R(t_j)$, is present only if there is observational error in the model.

6. Accumulate the quantities needed to calculate –2 ln likelihood at the end of the recursion,

$$\mathbf{M} \leftarrow \mathbf{M} + \mathbf{I}'(t_j)\mathbf{V}^{-1}(t_j)\mathbf{I}(t_j),$$

   the innovation variance is now a covariance matrix. The determinant term is now

$$\Delta \leftarrow \Delta + \ln|\mathbf{V}(t_j)|,$$

   where $|\mathbf{V}(t_j)|$ denotes the determinant of $\mathbf{V}(t_j)$.

7. To update the estimate of the state vector, let

$$\mathbf{A}(t_j) = \mathbf{H}(t_j)\mathbf{P}(t_j|t_j - \delta t),$$

   then

$$\mathbf{S}(t_j|t_j) = \mathbf{S}(t_j|t_j - \delta t) + \mathbf{A}'(t_j)\mathbf{V}^{-1}(t_j)\mathbf{I}(t_j).$$

8. The updated covariance matrix is,

$$\mathbf{P}(t_j|t_j) = \mathbf{P}(t_j|t_j - \delta t) - \mathbf{A}'(t_j)\mathbf{V}^{-1}(t_j)\mathbf{A}(t_j).$$

   In this step, $\mathbf{V}(t_j)$ is a matrix.

Now return to step 1 until the end of the data is reached.

At this point, the algorithm in Section 5.1.1 is followed to obtain the residual sum of squares (RSS). When a variance is not concentrated out of the likelihood, –2 ln likelihood is

$$\ell = n_T \ln(2\pi) + \Delta + RSS.$$

The n_T in this equation is the total number of observations. For example, in a balanced design without missing observations, if there are m subjects, n observation times for each subject, and d dependent variables for each subject, $n_T = mnd$. A sample program for these calculations is given in the Appendix on page 205.

8.3.1 Model fitting strategy

If the dimension of the multivariate response vector is d, there are d^2 elements in the **A** matrix, and $d(d+1)/2$ possible elements in each of the covariance matrices, the random input to the state equation, **G**, the observational error covariance matrix, **R**, and the covariance matrix of the between subject random effects, **U**. Here, the size of the covariance matrix of the random effects assumes that there is a single random effect for each element of the state vector which is simply a level shift. More complicated models for the between subject component of variance are possible. The richness of this error structure creates the possibility of non-identifiability of components in this error structure. A systematic approach is necessary to prevent a fishing expedition for error models.

It may be reasonable to restrict the observational error covariance matrix **R** to be diagonal. In many applications, this matrix will be a matrix of zeros. The recommended model fitting strategy is to fit univariate models first. This includes determining the independent variables to be included in the $\mathbf{X}_i$ matrices, as well as the error structure. A typical model may contain the elements A_{11} as the autoregression parameter, U_{11} as the standard deviation of the between subject random effect, G_{11} as the standard deviation of the random input to the state equation, and possibly $\sqrt{R_{11}}$, the standard deviation of the observational error. The inclusion of the observational error should be the last error parameter entered into the model since it is often not significant. Significant tests are based on changes in -2 ln likelihood.

When univariate models have been determined for each response variable, the situation is one of *seemingly unrelated regressions* (Srivastava and Giles, 1987). By moving to the multivariate models, the error structures are linked providing better estimates of the fixed effect parameters. When fitting the multivariate models, it is recommended that the fixed effects part of the model be left as it was in the univariate fits. The univariate error model parameters should be retained in the model and off diagonal elements, corresponding to existing diagonal elements, examined that relate to hypotheses being tested. A suggested order for including new parameters is to first look at off diagonal elements in the between subject random effects matrix **U**. Next look for significant off diagonal elements of the autoregression matrix **A**, and finally, off diagonal elements of the random input matrix, **G**.

8.4 A bivariate example

The San Luis Valley Diabetes Study (SLVDS) is a case control study of non-insulin-dependent diabetes mellitus (NIDDM) (type II diabetes, often referred to as adult onset diabetes) in the San Luis Valley in southern Colorado (Hamman et al., 1989). There are two major ethnic groups in the San Luis Valley, both of European ancestry. The Hispanics are mainly the descendants of the original Spanish settlers of the region who are often a European and American Indian admixture. The other ethnic group is referred to as non-Hispanics of European ancestry. These people migrated to the Valley largely from northern Europe. The American Indians have much higher rates of NIDDM than Europeans, and the Hispanics appear to have rates of NIDDM between the Europeans and American Indians.

The SLVDS attempted to locate all cases of NIDDM in a two county region of the San Luis Valley. A geographically based control group was selected with an age distribution similar to the cases. All cases and controls spent half a day in a clinic for an initial workup that included a glucose tolerance test. The subjects fasted for at least eight hours before arriving at the clinic. A blood sample was drawn and, among other tests, the fasting insulin level was determined. They ingested a dose of glucose followed by one hour and two hour blood drawings to determine the insulin and glucose levels. Based on these tests, some of the control subjects were determined to have NIDDM and were removed from the control group. Some subjects were classified as impaired glucose tolerance (IGT), a condition between normal and NIDDM. The IGT subjects were followed as a separate group with follow up visits to the clinic approximately every two years. At the time of this analysis, the subjects had between one and four visits to the clinic. The data consist of 173 subjects with a total of 504 clinic visits.

The bivariate response variables in the model are body mass index (BMI) which is weight in kilograms over height in meters squared, a measure of obesity, and $\log_{10}$ of the fasting insulin level. Other data available are the subject's age at the time of the initial clinic visit, the time in years for each clinic visit measured from the initial clinic visit, (t_{ij}), where the subscript i denotes the subject and j denotes the visit, and two indicator variables for the subject's gender (male=0, female=1) and ethnic group (non-hispanic=0, hispanic=1). Missing values are allowed in the two response variables but not in the independent variables. The questions to be considered are which of the variables contribute significantly to modelling the population characteristics, and what are the relationships between the response variables within subjects. If the 2×2 matrix $\mathbf{A}$ has all four elements non zero, the system would appear to be a feedback system with both variables affecting the other. Another possibility is that a third unmeasured variable is affecting both of the response variables; however, this may show up as an off diagonal term in the $\mathbf{G}$ matrix, the matrix multiplying the random input.

The most complete model considered, in the form of equation (8.16), is

$$\begin{bmatrix} y_{i1}(t_{ij}) \\ y_{i2}(t_{ij}) \end{bmatrix} = \begin{bmatrix} 1 & x_{i1}^{(1)} & \cdots & x_{i1}^{(p_1)} & 0 & 0 & 0 & 0 \\ 0 & 0 & 0 & 0 & 1 & x_{i2}^{(1)} & \cdots & x_{i2}^{(p_2)} \end{bmatrix} \boldsymbol{\beta}$$

$$+ \begin{bmatrix} 1 & 0 \\ 0 & 1 \end{bmatrix} \begin{bmatrix} \gamma_{i1} \\ \gamma_{i2} \end{bmatrix} + \begin{bmatrix} \epsilon_{i1}(t_{ij}) \\ \epsilon_{i2}(t_{ij}) \end{bmatrix}. \qquad (8.18)$$

The x's may be fixed or time varying covariates and may be a different set of variables for each dependent variable, and may or may not include the observation times, t_{ij}, and functions of the observation times such as powers or cosines. The random subject effects in the model are simply a level shift on each response variable for each subject. These level shifts for the two response variables can be correlated so the between subject covariance matrix, $\mathbf{B}$, is an arbitrary 2×2 matrix.

The fixed effect term containing the $\boldsymbol{\beta}$ can be concentrated out of the likelihood as in the univariate case except that the $\mathbf{X}_i$ matrix must be handled two rows at a time using the inverse of the 2×2 innovation covariance matrix as a weighting factor. The last two terms of (8.18) can be written in state space form. The state equation is

$$d \begin{bmatrix} \epsilon_{i1}(t) \\ \epsilon_{i2}(t) \\ \gamma_{i1} \\ \gamma_{i2} \end{bmatrix} = \begin{bmatrix} a_{11} & a_{12} & 0 & 0 \\ a_{21} & a_{22} & 0 & 0 \\ 0 & 0 & 0 & 0 \\ 0 & 0 & 0 & 0 \end{bmatrix} \begin{bmatrix} \epsilon_{i1}(t) \\ \epsilon_{i2}(t) \\ \gamma_{i1} \\ \gamma_{i2} \end{bmatrix} dt$$

$$+\begin{bmatrix} g_{11} & 0 \\ g_{21} & g_{22} \\ 0 & 0 \\ 0 & 0 \end{bmatrix} \begin{bmatrix} dw_{i1}(t) \\ dw_{i2}(t) \end{bmatrix},$$

and the observation equation is

$$\begin{bmatrix} \xi_{i1}(t_{ij}) \\ \xi_{i2}(t_{ij}) \end{bmatrix} = \begin{bmatrix} 1 & 0 & 1 & 0 \\ 0 & 1 & 0 & 1 \end{bmatrix} \begin{bmatrix} \epsilon_{i1}(t_{ij}) \\ \epsilon_{i2}(t_{ij}) \\ \gamma_{i1} \\ \gamma_{i2} \end{bmatrix} + \begin{bmatrix} v_{i1}(t_{ij}) \\ v_{i2}(t_{ij}) \end{bmatrix}.$$

If observations are missing, the corresponding rows of the observation equation are deleted. For example, if the first observation were missing, the observation equation in this example would be

$$[\xi_{i2}(t_{ij})] = [\begin{matrix} 0 & 1 & 0 & 1 \end{matrix}] \begin{bmatrix} \epsilon_{i1}(t_{ij}) \\ \epsilon_{i2}(t_{ij}) \\ \gamma_{i1} \\ \gamma_{i2} \end{bmatrix} + [v_{i2}(t_{ij})]$$

Fitting univariate models first, the model that minimized AIC for the response variable BMI is

```
Dependent variable:  BMI (kg/m**2)
```

	s.e.	t	
39.3	3.40	11.55	
-.186	.0568	-3.33	Age
1.31	.804	1.63	Sex
-8.78	4.42	-1.99	Ethnicity
.126	.0743	1.70	Age * Ethnicity

The estimates of the nonlinear parameters that give the minimum AIC are $U_{11} = 5.00$, $A_{11} = -.607$ and $G_{11} = 1.82$. -2 ln likelihood for this model is 2357.86, and AIC=2373.86 since there are five linear and three nonlinear parameters estimated.

The model that minimized AIC for the response variable log fasting insulin is

```
Dependent variable:  Log(10) Fasting Insulin

                s.e.        t
   1.09       .0294     36.99
-.00299       .0102     -0.29    Time (years)
  .0517       .0307      1.68    Sex
  .0607       .0300      2.02    Ethnicity
-.00340      .00182     -1.86    Time * Time
```

The estimates of the nonlinear parameters that give the minimum AIC are $U_{11} = .165$, $A_{11} = -.889$ and $G_{11} = .205$. −2 ln likelihood for this model is −278.02, and AIC=−262.02. In the two univariate models, the interaction terms are of marginal significance, and it was decided to leave them in the model. If the time squared term is deleted from the model, time becomes very significant.

When the two response variables are entered into the bivariate model with initial guesses at the nonlinear parameters as those obtained from the univariate models, and no additional parameters, the same results are obtained. This is to be expected since no coupling has been introduced into the model. The values of −2 ln likelihood and AIC obtained are the sum of the values from the two univariate fits, −2 ln likelihood=2079.84, and AIC= 2111.84.

The addition of the cross term in the random effects covariance matrix, U_{12} produces a very significant improvement in the model (−2 ln likelihood=2018.00, AIC=2052.00). This indicated that the random effects on the two response variables are positively correlated. If a subject is above the group mean on BMI, he/she will tend to be above the group mean on log fasting insulin. This observation is confirmed by the clinicians on the study.

Next, the two off diagonal elements of the **A** matrix are included in the model. The estimates from the previous model are used as initial guesses at the nonlinear parameters with the initial guesses at the new parameters being zero. By doing this, −2 ln likelihood at the initial guess is the final value obtained from the previous model. Since the minimum seeking routine does not find a value with a higher −2 ln likelihood, there is, hopefully, always improvement. The final result is that A_{21} is very significant, while A_{12} is not significant. Also, including G_{21} in the model causes no improvement. The estimates of the fixed effects for the final model are:

```
Dependent variable:  BMI (kg/m**2)
                 s.e.        t
   37.3          2.87      12.98
   -.153         .0467     -3.27      Age
   1.38          .804       1.12      Sex
   -7.96         3.72      -2.14      Ethnicity
   .113          .0621      1.81      Age * Ethnicity

Dependent variable:  Log(10) Fasting Insulin
                 s.e.        t
   1.09          .0294     36.99
 -.00101         .0097     -0.10      Time (years)
   .0531         .0308      1.73      Sex
   .0619         .0301      2.06      Ethnicity
 -.00355         .0017     -2.07      Time * Time
```

For this model, −2 ln likelihood is 1995.49, and AIC=2031.49. AIC will allow a variable into the model if the t-value is greater than the square root of 2 in absolute value. This means that the partial P-value is less than about 0.15. BMI shows a downward trend in the population with the subject's age, i.e. older subjects tend to have a lower BMI than younger subjects. There is also a downward trend in $\log_{10}$ fasting insulin with time of follow up. Since the origin of time for each subject is the time of the initial visit, this trend may be explained by a tendency of these IGT subjects to progress towards NIDDM.

The Cholesky factor of the matrix of the random subject effects and the covariance matrix of the random subject effects are estimated as

$$\mathbf{U} = \begin{bmatrix} 5.01 & .103 \\ 0 & .133 \end{bmatrix} \qquad \mathbf{B} = \mathbf{U}'\mathbf{U} = \begin{bmatrix} 25.1 & .516 \\ .516 & .0283 \end{bmatrix}.$$

The between subject standard deviation of BMI is estimated as 5.01 kilograms, and the between subject standard deviation of $\log_{10}$ fasting insulin is estimated as 0.168. Since this is on a log scale, a two standard deviation change in the log fasting insulin is equivalent to changing by a factor of 2.17 in the original insulin scale ($10^{2\times 0.168}$). The between subject correlation between BMI and log fasting insulin is estimated as .612.

The estimated **A** matrix is

$$\mathbf{A} = \begin{bmatrix} -.667 & 0 \\ .0599 & -1.258 \end{bmatrix}.$$

When **A** has this triangular form, the eigenvalues are the diagonal elements, $\lambda_1 = -.667$ and $\lambda_2 = -1.258$. The matrix of right eigenvectors of **A** is

$$\mathbf{C} = \begin{bmatrix} 1 & 0 \\ .1014 & 1 \end{bmatrix}$$

since $\mathbf{AC} = \mathbf{C\Lambda}$.

The matrix multiplying the random input to the state equation is estimated as

$$\mathbf{G} = \begin{bmatrix} 1.83 & 0 \\ 0 & .219 \end{bmatrix}.$$

No variances are concentrated out of the likelihood in order to allow any variance to go to zero. **K** can now be calculated from (8.13),

$$\mathbf{K} = \begin{bmatrix} 3.34 & -.338 \\ -.338 & .0822 \end{bmatrix},$$

and the covariance matrix of the random input to the rotated state equation, $\mathbf{Q}_r(\delta t)$, can be calculated from (8.14). The limit of $\mathbf{Q}_r$ for large δt gives the initial covariance matrix of the rotated state (8.15),

$$\mathbf{P}_r(0|0) = \begin{bmatrix} 2.50 & -.176 \\ -.176 & .0327 \end{bmatrix}.$$

The initial covariance matrix can then be rotated back to the original coordinate system,

$$\mathbf{P}(0|0) = \mathbf{CP}_r(0|0)\mathbf{C}^* = \begin{bmatrix} 2.50 & .0779 \\ .0779 & .0227 \end{bmatrix}.$$

This is the unconditional covariance matrix (lag zero) of the multivariate AR(1) process. The square roots of the diagonal elements give the estimated standard deviations of the elements of the multivariate process sd(BMI) = 1.58, sd(Log insulin) = .151, and the correlation between them, $\rho = .327$.

A tentative interpretation of the results is, because of the zero in the upper right hand corner of the **A** matrix, within subjects, the insulin level is affected by BMI and the insulin level, but BMI is affected only by BMI. This indicates that this may be an open loop system where BMI is causing the insulin level to change, and not a feedback system where both variables are affecting each other. This conclusion does not agree with the hypothesis of some of the investigators on the study who believe that there is feedback in the system. The investigators agree with the conclusion that there is a strong effect of BMI on fasting insulin. As BMI increases, the level of fasting insulin tends to increase. The hypothesized feedback is that increasing levels of fasting insulin should cause the rate of weight gain to decrease. This may be the case but we do not have long enough records on each subject to demonstrate the feedback effect.

Since the usual time between visits for subjects is about two years, it is interesting to see what the discrete time multivariate autoregression would be with a sampling interval of two years. From (8.5)

$$\mathbf{\Phi}(2) = \begin{bmatrix} 1 & 0 \\ .1014 & 1 \end{bmatrix} \begin{bmatrix} e^{(-.667\times 2)} & 0 \\ 0 & e^{(-1.258\times 2)} \end{bmatrix} \begin{bmatrix} 1 & 0 \\ .1014 & 1 \end{bmatrix}^{-1}$$

$$= \begin{bmatrix} .263 & 0 \\ .0185 & .0808 \end{bmatrix}. \tag{8.19}$$

This matrix also has a zero in the upper right hand corner indicating that BMI is the driving force. The covariance matrix of the random input to the state can be calculated for a time step of two years from (8.14) and (8.9),

$$\mathbf{Q}(2) = \begin{bmatrix} 2.33 & .0640 \\ .0640 & .0215 \end{bmatrix}. \tag{8.20}$$

8.4.1 Predictions

Since the within subject errors have a bivariate AR(1) structure, it is possible to make subject specific predictions. The state space approach makes predictions for individuals straight forward. Once the parameters of the model have been estimated using the data from all the subjects, the state space model can be used to make subject specific predictions for a given subject. These predictions can be used for such purposes as clinical monitoring of a subject. By predicting, with confidence intervals, where a subject should be at the next visit time, subjects who deviate significantly from the prediction can be noted for special consideration.

Consider a subject for which there are no observations. Data must be collected for the design matrices $\mathbf{X}_i$ and $\mathbf{Z}_i$ for this subject at the time that the first observation will be taken. The $\mathbf{Z}_i$ matrix will often be the identity matrix indicating a simple random shift in each component of the multivariate response variable. Each row of the $\mathbf{X}_i$ matrix may consist of a 1, possibly time and powers of time (or sines and cosines), possibly the subject's age, and indicator variables for such things as the subject's gender, or whether the subject is a case or a control. Given this baseline information, the prediction of the subject's first observation of the vector of response variables would be

$$\hat{\mathbf{y}}(t_{i1}) = \mathbf{X}_i(t_{i1})\hat{\boldsymbol{\beta}}.$$

At this time, there is no information about the values of the subject's random effects. The covariance of this 'prediction' is

$$\text{cov}\{\hat{\mathbf{y}}(t_{i1})\} = \mathbf{X}_i(t_{i1})\text{cov}(\boldsymbol{\beta})\mathbf{X}_i'(t_{i1}) + \sigma^2\mathbf{Z}_i(t_{i1})\mathbf{B}\mathbf{Z}_i'(t_{i1}) + \mathbf{R}.$$

If observations of the response variables are available for a subject, the Kalman filter is used, and run through the existing observations. From the point of the last observation, a prediction can be calculate for any step size by calculating $\mathbf{y}(t+\delta t|t)$, and the corresponding prediction covariance matrix $\mathbf{V}(t+\delta t)$. As more observations become available for a given subject, better estimates are obtained for that subject's random effects, $\boldsymbol{\gamma}_i$. This random effect vector will often be simply a random shift for each component of the response vector.

In discussions of people's weight, there has been much discussion of the concept of a 'set point'. The weight of a person who is in a steady state with a fairly constant activity level and diet will tend to fluctuate about some level. Equivalently, the person's BMI will tend to fluctuate about some level, since BMI is weight divided by height squared, and height does not tend to fluctuate. This set point will be partly determined by the person's covariables such as gender, age, etc. If other information is available, such as the person's activity level and information about diet, a more accurate determination of a person's set point can be determined. The part of a person's mean weight level that

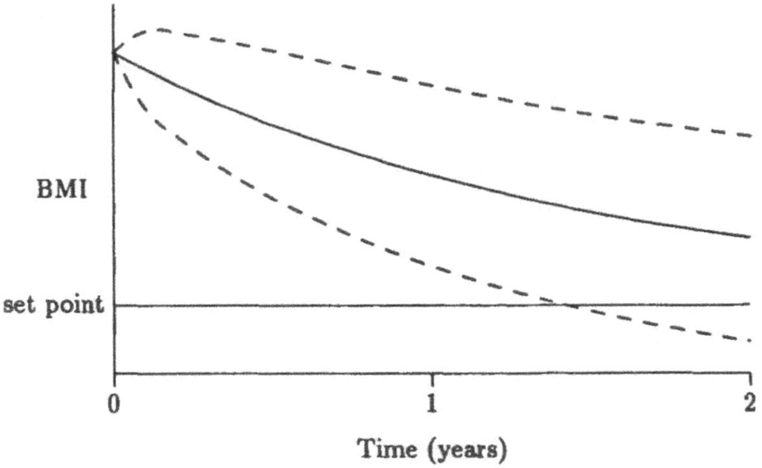

Fig. 8.1 *Prediction of a subject's BMI as a function of time with a high initial value. The dashed lines are plus and minus one standard deviation. In this model, BMI is not affected by the level of log fasting insulin.*

is not predictable from these covariates is modelled by the random effect, γ_i. With many observations on a given person, γ_i can be estimated, and therefore, the set point for this person can be estimated.

Fig. 8.1 shows predictions for a subject in the example in this section whose BMI (or weight) is above the set point at the time of the last observation. This subject's BMI is predicted to decay exponentially towards the set point, and the rate of decay is not affected by the subject's level of log insulin. Fig. 8.2 shows the prediction of this subject's log insulin when it starts above its set point with BMI at its set point. In this case, log insulin also decays exponentially towards its set point. Fig. 8.3 shows the prediction of the subject's log insulin when both log insulin and BMI start high. In this case, the rate of decay is slower than when BMI is not high. Fig. 8.4 shows the prediction of the subject's log insulin when log insulin starts at its set point, but BMI starts high. If both log insulin and BMI start at their set points, they would both be predicted to remain there. Since BMI starts high, it tends to pull the log insulin up. Fig. 8.5 shows the prediction of log insulin when log insulin starts low and BMI starts high. In

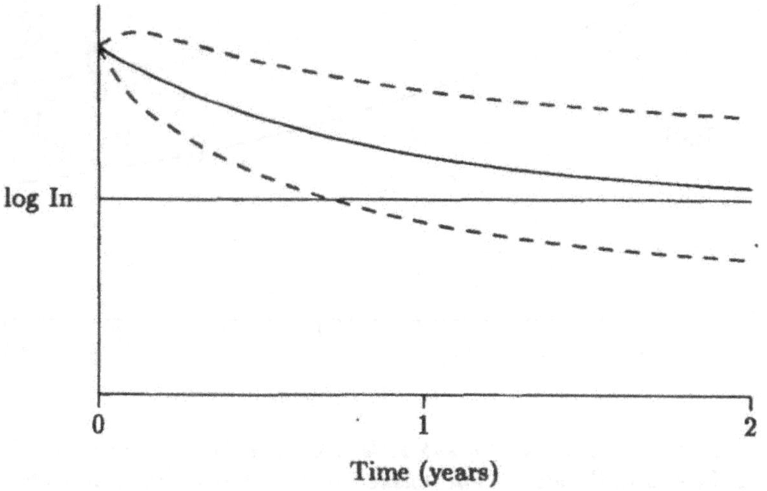

Fig. 8.2 *Prediction of a subject's log fasting insulin with BMI at the set point and log fasting insulin starting high.*

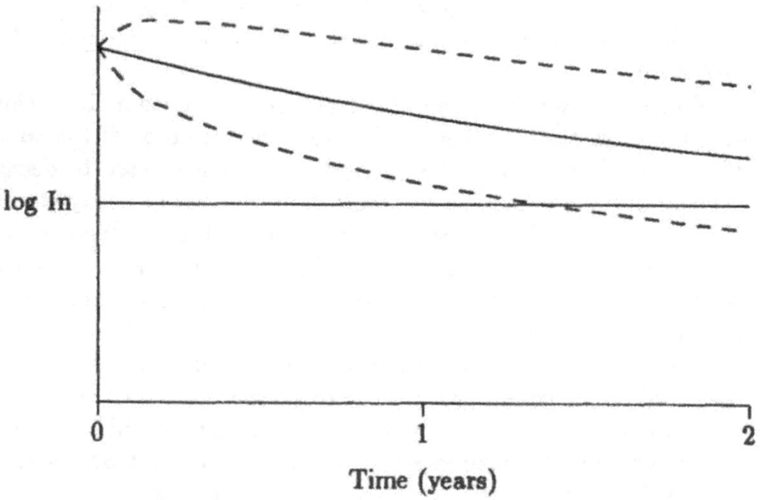

Fig. 8.3 *Prediction of a subject's log fasting insulin with BMI high and log fasting insulin starting high.*

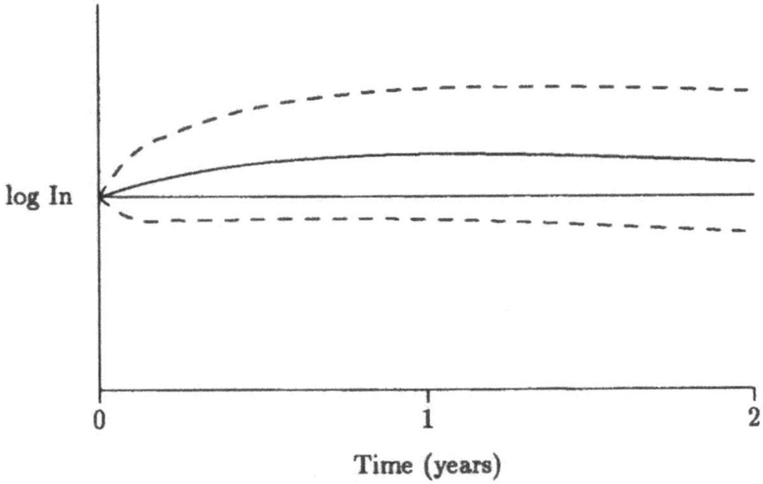

Fig. 8.4 *Prediction of a subject's log fasting insulin with BMI high and log fasting insulin starting at the set point.*

this case, instead of log insulin decaying towards the set point, it is pulled up over the set point by the high value of BMI.

These figures are meant to demonstrate a possible cause and effect relationship between two response variables when there is a zero in the **A** matrix. The level of the first variable can affect the rate of decay of the second variable, but the level of the second variable does not affect the rate of decay of the first variable. This is a direct interpretation of the continuous time state space models in equations (8.3) and (8.17). A discussion of causality between two vectors in multivariate ARMA models is given in Boudjellaba, Dufour and Roy (1992).

8.5 Doubly labeled water

Another example of a bivariate response variable is in the use of doubly labeled water to estimate a person's energy expenditure. Doubly labeled water is water containing stable isotopes of hydrogen (deuterium, ^2H), and oxygen, (^{18}O). The body eliminates hydrogen in the form of water, and oxygen in the form of both water and carbon dioxide (CO_2). Since CO_2 is a product of energy expenditure, after a dose of doubly labeled water, the difference

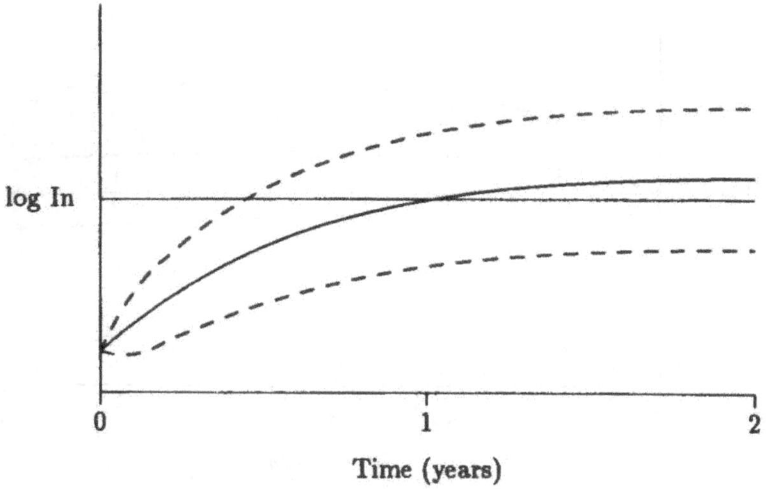

Fig. 8.5 *Prediction of a subject's log fasting insulin with BMI high and log fasting insulin starting low.*

in the rate of decay of ^{2}H and ^{18}O from the body depends on the energy expenditure of the subject. This rate of decay is determined from urine samples taken over a one or two week period following the dose. More details can be found in Jones (1993). A method for obtaining approximate confidence intervals on nonlinear parameters, similar to the method used by Jones and Molitoris (1984), has been used in this application. When the minimum of −2 ln likelihood is found, a search is carried out to determine how far a parameter needs to be moved from the minimum for −2 ln likelihood to increase significantly as chi-square with one degree of freedom. For each value of the parameter of interest tried in the search, −2 ln likelihood must be minimized with respect to the other parameters.

The basic model is two parallel single compartmental models (Jones, Reeve and Swanson, 1984). The continuous time state space equation is

$$d\begin{bmatrix} O(t) \\ H(t) \end{bmatrix} = \begin{bmatrix} -k_o & 0 \\ 0 & -k_h \end{bmatrix} \begin{bmatrix} O(t) \\ H(t) \end{bmatrix} dt + \begin{bmatrix} g_1 \\ g_2 \end{bmatrix} dw(t), \qquad (8.21)$$

and the observation equation for observations at time t_j is

$$\begin{bmatrix} y_o(t_j) \\ y_h(t_j) \end{bmatrix} = \begin{bmatrix} 1 & 0 \\ 0 & 1 \end{bmatrix} \begin{bmatrix} O(t_j) \\ H(t_j) \end{bmatrix} + \begin{bmatrix} v_o(t_j) \\ v_h(t_j) \end{bmatrix}.$$

In (8.21), $O(t)$ is the concentration of ^{18}O and $H(t)$ is the concentration of ^{2}H in the body water. The rates of decay of the concentrations are proportional to the concentrations. A single random input is included in the model which is distributed to the two components. It is this random input that introduces serial correlation into the model.

Since the state transition matrix is diagonal, no rotation of the state equation is necessary, and fitting the model simplifies greatly. If (8.21) is integrated over a finite time step of length δt, the resulting discrete time state equation is

$$\begin{bmatrix} O(t+\delta t) \\ H(t+\delta t) \end{bmatrix} = \begin{bmatrix} e^{-k_o \delta t} & 0 \\ 0 & e^{-k_h \delta t} \end{bmatrix} \begin{bmatrix} O(t) \\ H(t) \end{bmatrix} + \begin{bmatrix} u_1(t+\delta t) \\ u_2(t+\delta t) \end{bmatrix},$$

where the random input $\mathbf{u}(t+\delta t)$ has covariance matrix $\mathbf{Q}(\delta t)$ that can be calculated from g_1 and g_2 and depends on δt.

Data from three subjects are given in Prentice (1990, Chapter 11, pp. 218–220). The results of two fits to the data of Subject 1 are shown in Fig. 8.6 and 8.7. Fig. 8.6 is a fit with no random input to the state equation in the model, and Fig. 8.7 is a fit with random input in the model. Allowing random input to the state gives a much better fit to the data (AIC = −145.10) than no random input (AIC = −125.10), and the standardized residuals have a better pattern. It is clear from Fig. 8.7 that the residuals from ^{18}O and ^{2}H are correlated. This correlation is modelled by the random input to the state.

8.6 Exercise

1. For the example presented in this chapter, calculate the $\mathbf{\Phi}(1)$ and $\mathbf{Q}(1)$ matrices, similar to (8.19) and (8.20), for a prediction of one year.

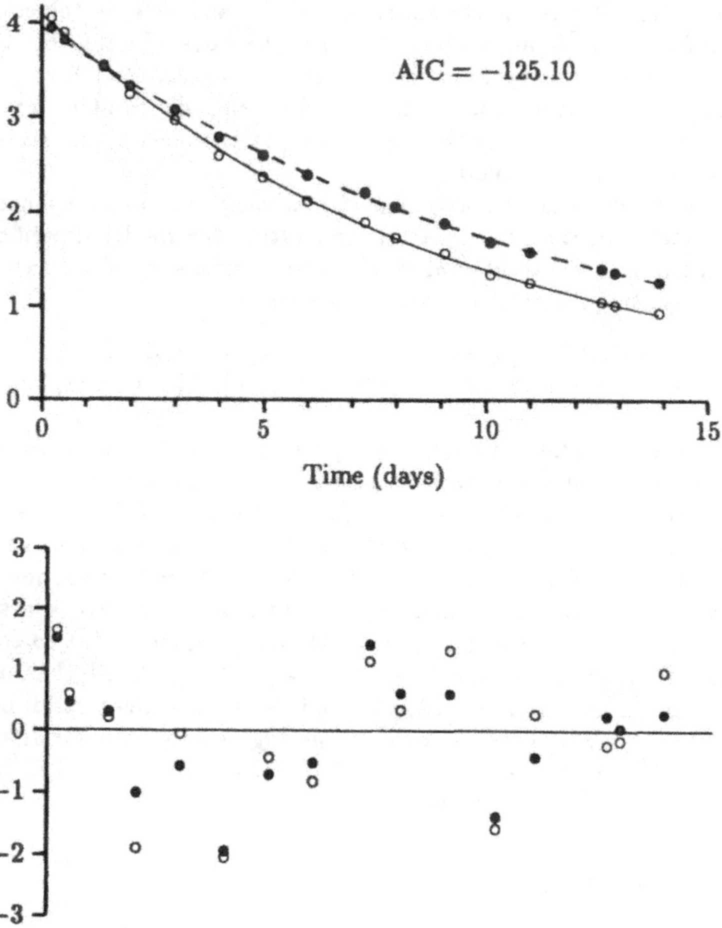

Fig. 8.6 ^{18}O (∘) and ^{2}H (•) concentrations and standardized residuals with no random input to state.

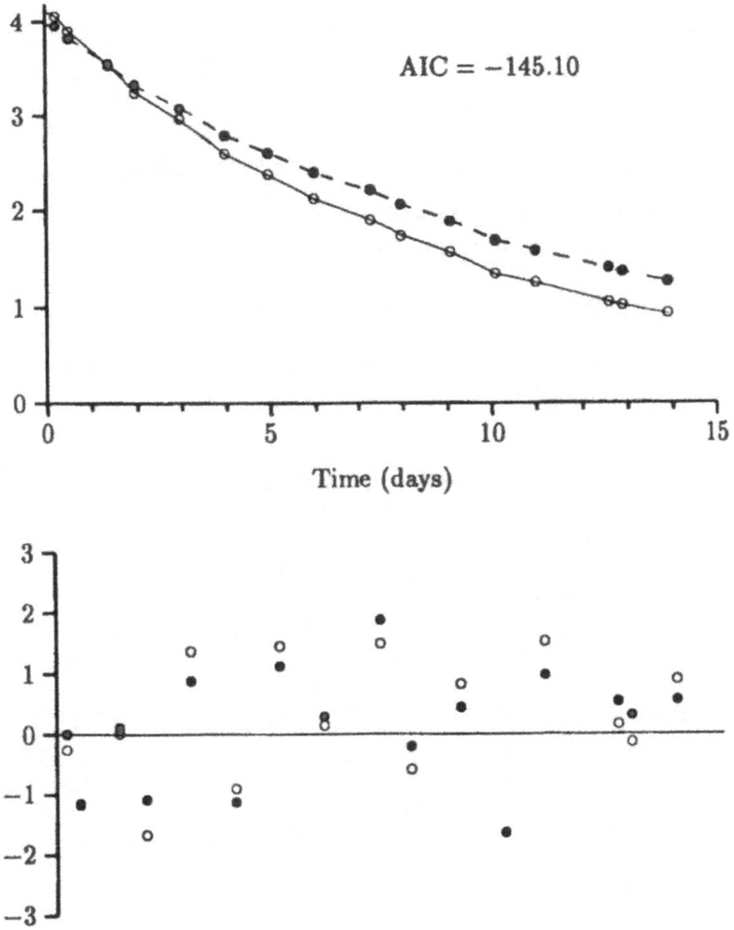

Fig. 8.7 ^{18}O (○) and ^{2}H (●) concentrations and standardized residuals with random input to state.

APPENDIX A

FORTRAN subroutines

A.1 Matrix factorization

The Cholesky factorization of an augmented matrix.

```
      SUBROUTINE FACTOR(V,N,ND,M)
      DOUBLE PRECISION V(ND,M)
C
C  THIS SUBROUTINE DOES AN IN PLACE UPPER TRIANGULAR CHOLESKY
C  FACTORIZATION OF A POSITIVE DEFINITE MATRIX, V=T'T.  THE
C  MATRIX MAY BE AUGMENTED BY ANY NUMBER OF COLUMNS WHICH
C  WILL BE REDUCED BY PREMULTIPLICATION BY T' INVERSE.
C
C  INPUT TO SUBROUTINE
C       V  POSITIVE DEFINITE N BY N MATRIX TO BE FACTORED.
C          ONLY THE UPPER TRIANGULAR PART IS USED AND IS
C          REPLACED BY T.  V MAY BE AUGMENTED BY M-N COLUMNS
C          TO BE PREMULTIPLIED BY T' INVERSE.
C       N  THE NUMBER OF ROWS OF V
C       ND  FIRST DIMENSION OF V IN THE CALLING PROGRAM
C       M  THE NUMBER OF COLUMNS OF V.  M MUST BE .GE. N
C
C  CHECK INPUT PARAMETERS
         IF(N.LT.1   .OR.
     1      N.GT.ND  .OR.
     2      M.LT.N       ) THEN
            WRITE(*,*) 'PARAMETERS OUT OF RANGE IN FACTOR'
            STOP
         END IF
C
      DO 60 I=1,N
         I1=I-1
         IF(I1.LT.1)GO TO 20
            DO 10 J=1,I1
               V(I,I)=V(I,I)-V(J,I)**2
   10       CONTINUE
```

```
20     CONTINUE
       IF(V(I,I).LE.0.0D0) GO TO 70
       V(I,I)=DSQRT(V(I,I))
       IP1=I+1
       IF(IP1.GT.M)RETURN
       DO 50 J=IP1,M
          IF(I1.LT.1)GO TO 40
             DO 30 K=1,I1
                V(I,J)=V(I,J)-V(K,I)*V(K,J)
30           CONTINUE
40        CONTINUE
          IF(V(I,I).LE.0.0D0) GO TO 70
          V(I,J)=V(I,J)/V(I,I)
50     CONTINUE
60  CONTINUE
    RETURN
70  WRITE(*,*) 'MATRIX NOT POSITIVE DEFINITE IN FACTOR'
    STOP
    END
```

After the matrix is factored, a call to the next subroutine solves the system or systems of equations when the right hand side of the systems of equations are in the augmented portion of the original matrix. This subroutine does not alter the triangular factor of the original matrix.

```
    SUBROUTINE BACK(T,N,ND,M)
    DOUBLE PRECISION T(ND,M)
    N1=N+1
    DO 40 J=N1,M
       DO 30 I=1,N
          II=N+1-I
          I1=II+1
          IF(I1.GT.N) GO TO 20
             DO 10 K=I1,N
                T(II,J)=T(II,J)-T(II,K)*T(K,J)
10           CONTINUE
20        CONTINUE
       IF(T(II,II).LE.0.0D0) THEN
          T(II,J)=0.0D0
       ELSE
          T(II,J)=T(II,J)/T(II,II)
       END IF
30     CONTINUE
40  CONTINUE
```

```
      RETURN
      END
```

The following subroutine does an in place calculation of the inverse of the original matrix.

```
      SUBROUTINE TTVERT(T,N,ND)
      DOUBLE PRECISION T(ND,N),EX
C
C  THIS SUBROUTINE REPLACES THE UPPER TRIANGULAR FACTOR T BY
C  T'T INVERSE.  REF: GRAYBILL (1976) P. 246.
C
C  INPUT TO SUBROUTINE
C       T   UPPER TRIANGULAR N BY N MATRIX
C       ND  FIRST DIMENSION OF T IN CALLING PROGRAM
C
C  OUTPUT FROM SUBROUTINE
C       T   T'T INVERSE WHICH IS SYMMETRIC
      DO 50 JJ=1,N
         J=N+1-JJ
C  CALCULATE DIAGONAL ELEMENTS
         IF(T(J,J).LE.0.0D0) THEN
            T(J,J)=0.0D0
         ELSE
         EX=1.0D0/T(J,J)
         IF(J.LT.N)THEN
            DO 10 K=J+1,N
               EX=EX-T(J,K)*T(K,J)
   10       CONTINUE
         END IF
         T(J,J)=EX/T(J,J)
         END IF
C  FILL IN ROW J FROM TRANSPOSED POSITION
         IF(J.LT.N)THEN
            DO 20 K=J+1,N
               T(J,K)=T(K,J)
   20       CONTINUE
         END IF
C  CALCULATE COLUMN J AND STORE IN TRANSPOSED POSITION
         IF(J.EQ.1)RETURN
         DO 40 II=1,J-1
            I=J-II
            IF(T(I,I).GT.0.0D0) THEN
            EX=0.0D0
            DO 30 K=I+1,N
```

```
              EX=EX+T(I,K)*T(J,K)
   30       CONTINUE
            T(J,I)=-EX/T(I,I)
            ELSE
              T(J,I)=0.0D0
            END IF
   40     CONTINUE
   50 CONTINUE
      RETURN
      END
```

A.2 Equally spaced ARMA processes

The first subroutine untransforms the parameters of the model. Transformed parameters are used to keep the coefficients in the proper range.

```
      SUBROUTINE UNTRAN(NP,NQ,NE,P,Q)
      REAL P(11),Q(11),A(6)
C THIS SUBROUTINE CALCULATES THE AUTOREGRESSIVE COEFFICIENTS
C MOVING AVERAGE COEFFICIENTS AND OBSERVATIONAL ERROR
C VARIANCE FROM THE TRANSFORMED PARTIAL COEFFICIENTS AND
C ERROR STANDARD DEVIATION
C
C INPUT TO SUBROUTINE
C     NP--NUMBER OF AUTOREGRESSION COEFFICIENTS
C     NQ--NUMBER OF MOVING AVERAGE COEFFICIENTS
C     NE=0  NO OBSERVATIONAL ERROR
C       =1  OBSERVATIONAL ERROR
C     P--TRANSFORMED VALUES OF PARTIAL AUTOREGRESSIVE
C        COEFFICIENTS, MOVING AVERAGE COEFFICIENTS AND
C        OBSERVATIONAL ERROR STANDARD DEVIATION
C
C OUTPUT FROM SUBROUTINE
C     Q--AUTOREGRESSIVE COEFFICIENTS, MOVING AVERAGE
C        COEFFICIENTS AND OBSERVATIONAL ERROR VARIANCE
C
C UNTRANSFORM AUTOREGRESSIVE PART
      IF(NP.GT.0) THEN
        DO 30 I=1,NP
          EX=EXP(-P(I))
          Q(I)=(1.-EX)/(1.+EX)
          A(I)=Q(I)
          IF(I.GT.1) THEN
```

```
            DO 10 K=1,I-1
              Q(K)=A(K)-Q(I)*A(I-K)
 10         CONTINUE
            DO 20 K=1,I-1
              A(K)=Q(K)
 20         CONTINUE
          END IF
 30     CONTINUE
      END IF
C  UNTRANSFORM MOVING AVERAGE PART
      IF(NQ.GT.0) THEN
        DO 60 I=1,NQ
          EX=EXP(-P(NP+I))
          Q(NP+I)=(1.-EX)/(1.+EX)
          A(I)=Q(NP+I)
          IF(I.GT.1) THEN
            DO 40 K=1,I-1
              Q(NP+K)=A(K)+Q(NP+I)*A(I-K)
 40         CONTINUE
            DO 50 K=1,I-1
              A(K)=Q(NP+K)
 50         CONTINUE
          END IF
 60     CONTINUE
      END IF
C  UNTRANSFORM OBSERVATIONAL ERROR
      IF(NE.EQ.1) Q(NP+NQ+NE)=P(NP+NQ+NE)**2
      RETURN
      END
```

The inverse program that takes autoregressive coefficients, moving average coefficients and the observational error variance and transforms them is not actually necessary for the program since it works only with transformed variables. However, if actual coefficients are used rather than transformed parameters, it is necessary to transform them. An interesting application of the subroutine is to determine if a set of autoregression coefficients generate a stationary process. If this is not the case, an error message is printed out.

```
      SUBROUTINE TRANS(NP,NQ,NE,Q,P,IER)
      REAL Q(11),P(11)
C
C THIS SUBROUTINE TRANSFORMS THE AUTOREGRESSIVE COEFFICIENTS
C MOVING AVERAGE COEFFICIENTS AND OBSERVATIONAL ERROR
```

```
C VARIANCE TO THE TRANSFORMED PARTIAL COEFFICIENTS AND ERROR
C STANDARD DEVIATIONS
C
C INPUT TO SUBROUTINE
C     NP : NUMBER OF AUTOREGRESSION COEFFICIENTS
C     NQ : NUMBER OF MOVING AVERAGE COEFFICIENTS
C     NE = 0  NO OBSERVATIONAL ERROR
C        = 1  OBSERVATIONAL ERROR
C     Q  : AUTOREGRESSIVE COEFFICIENTS, MOVING AVERAGE
C          COEFFICIENTS AND OBSERVATIONAL ERROR VARIANCE
C
C  OUTPUT FROM SUBROUTINE
C     P : TRANSFORMED VALUES OF PARTIAL AUTOREGRESSIVE
C         COEFFICIENTS, MOVING AVERAGE COEFFICIENTS AND
C         OBSERVATIONAL ERROR STANDARD DEVIATION
C     IER : 0 => OK; 1 => BAD AR PARAMETERS; 2 => BAD MA
C         PARAMETERS
C
C TRANSFORM AUTOREGRESSIVE PART
C
      IER = 0
      IF(NP.GT.0)THEN
         DO 10 I=1,NP
            P(I)=Q(I)
   10    CONTINUE
         DO 40 IK=1,NP
            N=NP-IK+1
            PS=P(N)**2
            IF(PS.GE.1.0)THEN
               IER = 1
               RETURN
            END IF
            IF(N.EQ.1)GO TO 30
            D=1.0-PS
            N2=N/2
            DO 20 I=1,N2
               IF(N-I.GT.I)THEN
                  EX=(P(I)+P(N-I)*P(N))/D
                  P(N-I)=(P(N-I)+P(I)*P(N))/D
                  P(I)=EX
               ELSE
                  P(I)=P(I)/(1.0-P(N))
               END IF
   20       CONTINUE
```

```
   30       CONTINUE
            P(N)=ALOG((1.0+P(N))/(1.0-P(N)))
   40    CONTINUE
      END IF
C
C  TRANSFORM MOVING AVERAGE PART
C
      IF(NQ.GT.0)THEN
         DO 50 I=1,NQ
            P(NP+I)=Q(NP+I)
   50    CONTINUE
         DO 80 IK=1,NQ
            N=NQ-IK+1
            PS=P(NP+N)**2
            IF(PS.GE.1.0)THEN
               IER = 2
               RETURN
            END IF
            IF(N.EQ.1)GO TO 70
            D=1.0-PS
            N2=N/2
            DO 60 I=1,N2
               IF(N-I.GT.I)THEN
                  EX=(P(NP+I)-P(NP+N-I)*P(NP+N))/D
                  P(NP+N-I)=(P(NP+N-I)-P(NP+I)*P(N+NP))/D
                  P(NP+I)=EX
               ELSE
                  P(NP+I)=P(NP+I)/(1.0+P(NP+N))
               END IF
   60       CONTINUE
   70       CONTINUE
            P(NP+N)=ALOG((1.0+P(NP+N))/(1.0-P(NP+N)))
   80    CONTINUE
      END IF
C
C  TRANSFORM OBSERVATIONAL ERROR VARIANCE
C
      IF(NE.EQ.1)P(NP+NQ+NE)=SQRT(ABS(Q(NP+NQ+NE)))
      RETURN
  100 WRITE(6,1000)
 1000 FORMAT(' AUTOREGRESSIVE COEFFICIENTS NOT INVERTIBLE')
      RETURN
  200 WRITE(6,2000)
 2000 FORMAT(' MOVING AVERAGE COEFFICIENTS NOT INVERTIBLE')
```

```
      RETURN
      END
```

The next subroutine calculates the ψ weights or impulse response function. It assumes that the autoregression coefficients followed by the moving average coefficients are in the single precision array Q.

```
      SUBROUTINE CROSS(NP,NQ,Q,PSI)
      REAL Q(11)
      DOUBLE PRECISION PSI(6)
C  THIS SUBROUTINE CALCULATES THE IMPULSE RESPONSE FUNCTION
C  (PSI WEIGHTS) OF AN ARMA PROCESS
C
C  INPUT TO SUBROUTINE
C     NP--NUMBER OF AUTOREGRESSION COEFFICIENTS
C     NQ--NUMBER OF MOVING AVERAGE COEFFICIENTS
C     Q--VECTOR OF AUTOREGRESSION AND MOVING AVERAGE
C        COEFFICIENTS
C
C  OUTPUT FROM SUBROUTINE
C     PSI--FIRST MAX0(NP,NQ+1) VALUES OF IMPULSE RESPONSE
C          FUNCTION
C
      M=MAX0(NP,NQ+1)
      PSI(1)=1.0
      IF(M.GT.1) THEN
C  MOVING AVERAGE CONTRIBUTION
         DO 20 I=2,M
            IF(NQ.LT.I-1) THEN
               PSI(I)=0.0
            ELSE
               PSI(I)=Q(NP+I-1)
            END IF
C  AUTOREGRESSIVE CONTRIBUTION
            IF(NP.GT.0)THEN
               NPM=MIN0(NP,I-1)
               DO 10 J=1,NPM
                  PSI(I)=PSI(I)+Q(J)*PSI(I-J)
   10          CONTINUE
            END IF
   20    CONTINUE
      END IF
      RETURN
      END
```

The next subroutine uses the output from CROSS and calculates the correlation function of an ARMA(p,q) process. The LU (Lower triangular–Upper triangular) decomposition method (Press, et al., 1986, p. 31) is used to solve the system of linear equations.

```
      SUBROUTINE CORR(NP,NQ,MLAG,Q,PSI,C,RATIO)
      REAL Q(11)
      DOUBLE PRECISION PSI(6),C(MLAG+1),COEF(7,7),RATIO
      INTEGER INDX(7)
C THIS SUBROUTINE CALCULATES THE CORRELATION FUNCTION OF AN
C ARMA(NP,NQ) PROCESS TO LAG MAXO(NP,NQ,MLAG)
C
C INPUT TO SUBROUTINE
C      NP--NUMBER OF AUTOREGRESSION COEFFICIENTS
C      NQ--NUMBER OF MOVING AVERAGE COEFFICIENTS
C      MLAG--MAXIMUM LAG IF GREATER THAN NP AND NQ
C      Q--ARRAY OF AUTOREGRESSION AND MOVING AVERAGE
C         COEFFICIENTS
C      PSI--IMPULSE RESPONSE FUNCTION FROM SUBROUTINE CROSS
C
C OUTPUT FROM SUBROUTINE
C      C--CORRELATION FUNCTION ARRAY
C      RATIO--PROCESS VARIANCE / INPUT NOISE VARIANCE
C
C INITIALIZE COEFFICIENT MATRIX AND RIGHT HAND SIDE VECTOR
      DO 20 I=1,NP+1
         C(I)=0.0
         DO 10 J=1,NP+1
            COEF(I,J)=0.0
   10    CONTINUE
         COEF(I,I)=1.0
   20 CONTINUE
C  FILL IN RIGHT HAND SIDE
      C(1)=1.0
      IF(NQ.GT.0) THEN
         DO 30 J=1,NQ
            C(1)=C(1)+Q(NP+J)*PSI(J+1)
   30    CONTINUE
         MINPQ=MINO(NP,NQ)
         IF(MINPQ.GT.0) THEN
            DO 50 I=1,MINPQ
               DO 40 J=I,NQ
                  C(I+1)=C(I+1)+Q(NP+J)*PSI(J-I+1)
   40          CONTINUE
```

```
   50       CONTINUE
          END IF
        END IF
C   FILL IN COEFFICIENT MATRIX
        IF(NP.GT.0)THEN
          DO 70 I=1,NP+1
            DO 60 K=1,NP
              J=1+IABS(I-1-K)
              COEF(I,J)=COEF(I,J)-Q(K)
   60       CONTINUE
   70     CONTINUE
        END IF
C   SOLVE SYSTEM OF EQUATIONS USING NUMERICAL RECIPES ROUTINE
        CALL LUDCMP(COEF,NP+1,7,INDX,D)
        CALL LUBKSB(COEF,NP+1,7,INDX,C)
C   CONTINUE COVARIANCE FUNCTION TO LAG NQ IF NECESSARY
        LAG=NP
        IF(NQ.GT.NP) THEN
          DO 100 I=NP+1,NQ
            C(I+1)=0.0
            IF(NP.GT.0) THEN
              DO 80 K=1,NP
                C(I+1)=C(I+1)+Q(K)*C(I+1-K)
   80         CONTINUE
            END IF
            DO 90 K=I,NQ
              C(I+1)=C(I+1)+Q(NP+K)*PSI(K-I+1)
   90       CONTINUE
            LAG=I
  100     CONTINUE
        END IF
        IF(MLAG.GT.LAG)THEN
          DO 120 I=LAG+1,MLAG
            C(I+1)=0.0
            IF(NP.GT.0)THEN
              DO 110 K=1,NP
                C(I+1)=C(I+1)+Q(K)*C(I+1-K)
  110         CONTINUE
            END IF
  120     CONTINUE
        END IF
        RATIO=C(1)
        DO 130 I=1,MAX0(NP,NQ,MLAG)+1
          C(I)=C(I)/RATIO
```

```
  130 CONTINUE
      RETURN
      END
```

This subroutine can be used to calculate the correlations of an ARMA(p,q) process at any lag. Therefore, it can be used to calculate the within subject correlation matrix $\mathbf{W}_i$ for use in the direct algorithm (section 2.2) for any pattern of missing observations.

The next subroutine calculates the initial state covariance matrix. Since this subroutine calls CORR, it only requires output from subroutine CROSS.

```
      SUBROUTINE ICOV(NP,NQ,Q,PSI,COV,RATIO)
      REAL Q(11)
      DOUBLE PRECISION PSI(6),COV(6,6),C(7),RATIO
C THIS SUBROUTINE INITIALIZES THE COVARIANCE MATRIX OF THE
C STATE VECTOR FOR AKAIKE'S MARKOVIAN REPRESENTATION BASED
C ON THE INPUT AUTOREGRESSIVE AND MOVING AVERAGE PARAMETERS
C
C INPUT TO SUBROUTINE
C     Q--VECTOR OF NP AUTOREGRESSIVE COEFFICIENTS FOLLOWED
C        BY NQ MOVING AVERAGE COEFFICIENTS
C     PSI--IMPULSE RESPONSE FUNCTION FROM SUBROUTINE CROSS
C
C OUTPUT FROM SUBROUTINE
C     COV--MAXO(NP,NQ+1) BY MAXO(NP,NQ+1) INITIAL STATE
C          COVARIANCE MATRIX NORMALIZED SO THAT COV(1,1)=1
C     C--CORRELATION FUNCTION FOR LAGS 0 TO MAXO(NP,NQ)
C     RATIO--THE PROCESS VARIANCE/RANDOM INPUT VARIANCE RATIO
C
      M=MAXO(NP,NQ+1)
C  CALCULATE COVARIANCES OF TIME SERIES
      CALL CORR(NP,NQ,MLAG,Q,PSI,C,RATIO)
      DO 30 I=1,M
         DO 20 J=I,M
            COV(I,J)=C(J-I+1)
            IF(I.GT.1) THEN
               DO 10 K=1,I-1
                  COV(I,J)=COV(I,J)-PSI(K)*PSI(K+J-I)
   10          CONTINUE
            END IF
            COV(I,J)=COV(I,J)
            COV(J,I)=COV(I,J)
   20    CONTINUE
   30 CONTINUE
```

```
      RETURN
      END
```

A.3 Unequally spaced ARMA processes

The first subroutine calculates the roots of the characteristic equation from the logs of the a's in the factored form of the characteristic equation (6.10). Note the necessity for complex arithmetic since some of the roots may be complex conjugate pairs.

```
      SUBROUTINE ROOTS(NAR,P,R,RES)
      PARAMETER(MAXAR=5)
      REAL P(MAXAR)
      DOUBLE PRECISION RE,A,FAC,ROOT
      COMPLEX*16 R(MAXAR),RES(MAXAR)
      DO 10 I=1,NAR,2
         IF(I.LT.NAR)THEN
            RE=EXP(P(I+1))/2.0D0
            A=EXP(P(I))
            FAC=RE*RE-A
            IF(FAC.EQ.0.0)THEN
               PRINT 1000
 1000    FORMAT(' DOUBLE ROOT, TRY NEW INITIAL ESTIMATES')
               STOP
            ELSE
               ROOT=DSQRT(DABS(FAC))
               IF(FAC.LT.0.0)THEN
                  R(I)=DCMPLX(-RE,-ROOT)
                  R(I+1)=DCMPLX(-RE,ROOT)
               ELSE
                  R(I)=-RE-ROOT
                  R(I+1)=-RE+ROOT
               END IF
            END IF
         ELSE
            R(I)=-DEXP(P(I))
         END IF
   10 CONTINUE
C
C  CALCULATE DENOMINATORS FOR COVARIANCE FUNCTION
C
      DO 30 J=1,NAR
         RES(J)=-2*REAL(R(J))
         DO 20 K=1,NAR
```

```
      IF(K.NE.J)RES(J)=RES(J)*(R(K)-R(J))*(DCONJG(R(K))+R(J))
   20    CONTINUE
   30 CONTINUE
      RETURN
      END
```

The next subroutine calculates the moving average parameters, the δ's, from the square roots of the coefficients in equation (6.11).

```
      SUBROUTINE UNMA(NAR,NMA,P,PMA)
      PARAMETER(MAXAR=5,MAXMA=4)
      REAL P(NAR+NMA)
      DOUBLE PRECISION PMA(NMA),TEMP(MAXAR+MAXMA)
C
C THIS SUBROUTINE UNTRANSFORMS THE MOVING AVERAGE
C COEFFICIENTS. THE INPUT COEFFICIENTS ARE SQUARED TO GIVE
C NONNEGATIVE COEFFICIENTS THAT ARE COEFFICIENTS IN
C QUADRATIC FACTORS WITH A TERMINATING LINEAR FACTOR IF NMA
C IS ODD.  THE POLYNOMIAL FACTORS ARE:
C
C (1 + (P(NAR+1)**2)*Z + (P(NAR+2)**2)*Z**2)
C (1 + (P(NAR+3)**2)*Z + (P(NAR+4)**2)*Z**2)
C    .
C    .
C    .
C (1+(P(NAR+NMA-1)**2)*Z+(P(NAR+NMA)**2)*Z**2) IF NMA IS EVEN
C (1+(P(NAR+NMA)**2)*Z)                        IF NMA IS ODD
C
C THE NMA COEFFICIENTS OF THE POLYNOMIAL OBTAINED BY
C MULTIPLYING THE FACTORS ARE CALCULATED AND STORED IN PMA
C
      DO 10 I=1,NMA
         PMA(I)=P(NAR+I)**2
   10 CONTINUE
      IF(NMA.LE.2)RETURN
      DO 50 I=3,NMA,2
         TEMP(1)=PMA(I)
         DO 20 J=2,I
            TEMP(J)=PMA(J-1)*PMA(I)
   20    CONTINUE
         IF(I.LT.NMA)THEN
            TEMP(2)=TEMP(2)+PMA(I+1)
            DO 30 J=3,I
               TEMP(J)=TEMP(J)+PMA(J-2)*PMA(I+1)
```

```
30         CONTINUE
           PMA(I+1)=PMA(I-1)*PMA(I+1)
        END IF
        DO 40 J=1,I-1
           PMA(J)=PMA(J)+TEMP(J)
40      CONTINUE
        PMA(I)=TEMP(I)
50 CONTINUE
   RETURN
   END
```

The covariance function of a continuous time ARMA process can be calculated up to a multiplicative constant by assuming that $\sigma_\eta^2 = 1$. This scaled covariance function can be normalized to a correlation function by calculating the scaled covariance at lag zero and then dividing the covariances by this number. This number must be saved since it is the ratio of the process variance to the input random noise variance. When using the correlation function in the analysis, the variance of the process is estimated by the mean square error. The next subroutine is a specialized version of the subroutine to calculate the covariance which follows, as it calculates the covariance at lag zero. It need only be called once for given values of the autoregressive and moving average parameters.

```
      SUBROUTINE VAR(NAR,NMA,PMA,RES,R,CO)
      PARAMETER(MAXAR=5,MAXMA=4)
      DOUBLE PRECISION CO,PMA(MAXMA)
      COMPLEX*16 R(MAXAR),RES(MAXAR),B1,B2,V
      V=(0.0D0,0.0D0)
      IF(NMA.EQ.0)THEN
         DO 10 K=1,NAR
            V=V+(1.0D0,0.0D0)/RES(K)
10       CONTINUE
      ELSE
         DO 30 K=1,NAR
            B1=(1.0D0,0.0D0)
            B2=(1.0D0,0.0D0)
            ISIGN=1
            DO 20 L=1,NMA
               ISIGN=-ISIGN
               B1=B1 + PMA(L)*R(K)**L
               B2=B2+ISIGN*PMA(L)*R(K)**L
20          CONTINUE
            V=V+B1*B2/RES(K)
```

```
30       CONTINUE
      END IF
      CO=V
      RETURN
      END
```

The next subroutine calculates the covariance function assuming unit random noise variance and normalizes it to a correlation function. This can be used to calculate the $\mathbf{W}_i$ matrices for the direct method of evaluating the likelihood, page 32.

```
      SUBROUTINE CORR(NAR,NMA,PMA,RES,R,LAG,CO,C)
      PARAMETER(MAXAR=5,MAXMA=4)
      REAL LAG
      DOUBLE PRECISION CO,C,PMA(MAXMA)
      COMPLEX*16 R(MAXAR),RES(MAXAR),B1,B2,V
      V=(0.0D0,0.0D0)
      IF(NMA.EQ.0)THEN
         DO 10 K=1,NAR
            V=V+CDEXP(R(K)*LAG)/RES(K)
10       CONTINUE
      ELSE
         DO 30 K=1,NAR
            B1=(1.0D0,0.0D0)
            B2=(1.0D0,0.0D0)
            ISIGN=1
            DO 20 L=1,NMA
               ISIGN=-ISIGN
               B1=B1 + PMA(L)*R(K)**L
               B2=B2+ISIGN*PMA(L)*R(K)**L
20          CONTINUE
            V=V+B1*B2*CDEXP(R(K)*LAG)/RES(K)
30       CONTINUE
      END IF
      C=V/CO
      RETURN
      END
```

For the state space approach, to calculate the upper left hand block of the initial state covariance matrix for the rotated state in equation (6.23), it is first necessary to calculate the initial state covariance matrix from equation (6.21), and rotate it (6.22). This subroutine calculates the unrotated initial state covariance matrix and factors it using the Cholesky factorization routine FACTOR given on page 186.

```
      SUBROUTINE INIT(NAR,R,RES,RCOV,VARI)
      PARAMETER(MAXAR=5)
      DOUBLE PRECISION RCOV(MAXAR,MAXAR),VARI
      COMPLEX*16 R(MAXAR),RES(MAXAR),EX
C
C CALCULATE REAL COVARIANCE MATRIX OF STATE VECTOR
C SCALE SO UPPER LEFT HAND CORNER, THE VARIANCE OF THE
C PROCESS IS 1.  R AND RES ARE OUTPUT FROM SUBROUTINE ROOTS.
C THE OUTPUT COVARIANCE IS RCOV AND VARI IS THE RATIO OF THE
C PROCESS VARIANCE TO THE INPUT NOISE VARIANCE.
C
      DO 40 K=1,NAR
         DO 35 L=K,NAR
            EX=(0.0,0.0)
            DO 30 J=1,NAR
               EX=EX+(R(J)**(K-1))*((-R(J))**(L-1))/RES(J)
   30       CONTINUE
            RCOV(K,L)=EX
            IF(L.EQ.1)VARI=EX
            RCOV(K,L)=RCOV(K,L)/VARI
   35    CONTINUE
   40 CONTINUE
      CALL FACTOR(RCOV,NAR,MAXAR,NAR)
      RETURN
      END
```

The matrix of eigenvectors in equation (6.18) used for the rotation is calculated from the following subroutine using the roots in array R from subroutine ROOTS, page 197.

```
      SUBROUTINE EIGENV(NAR,R,C)
      PARAMETER(MAXAR=5)
      COMPLEX*16 R(MAXAR),C(MAXAR,MAXAR)
      DO 20 J=1,NAR
         C(1,J)=(1.0D0,0.0D0)
         IF(NAR.GT.1)THEN
            DO 10 I=2,NAR
               C(I,J)=C(I-1,J)*R(J)
   10       CONTINUE
         END IF
   20 CONTINUE
      RETURN
      END
```

The inverse of this rotation matrix appears in equations (6.19) and (6.22), but it is not necessary to actually calculate the inverse.

The necessary solutions can be calculated by solving systems of complex linear equations. This can be carried out by using a complex version of the LU (Lower triangular–Upper triangular) decomposition method (Press, et al., 1986, p. 31). This subroutine actually produces an LU decomposition of a rowwise permutation of the matrix A, but this is allowed for in the algorithm that follows.

```
      SUBROUTINE CLU(A,N,NP,INDX,D)
      PARAMETER (NMAX=100)
      INTEGER INDX(N)
      DOUBLE PRECISION AAMAX,DUM,VV(NMAX),TINY
      COMPLEX*16 A(NP,NP),SUM,EX
      D=1.
      DO 12 I=1,N
        AAMAX=0.0D0
        DO 11 J=1,N
          IF (CDABS(A(I,J)).GT.AAMAX) AAMAX=CDABS(A(I,J))
11      CONTINUE
        IF (AAMAX.EQ.(0.0D0,0.0D0)) PAUSE 'SINGULAR MATRIX.'
        VV(I)=1./AAMAX
12    CONTINUE
      DO 19 J=1,N
        IF (J.GT.1) THEN
          DO 14 I=1,J-1
            SUM=A(I,J)
            IF (I.GT.1)THEN
              DO 13 K=1,I-1
                SUM=SUM-A(I,K)*A(K,J)
13            CONTINUE
              A(I,J)=SUM
            END IF
14        CONTINUE
        END IF
        AAMAX=0.
        DO 16 I=J,N
          SUM=A(I,J)
          IF (J.GT.1)THEN
            DO 15 K=1,J-1
              SUM=SUM-A(I,K)*A(K,J)
15          CONTINUE
            A(I,J)=SUM
          END IF
          DUM=VV(I)*CDABS(SUM)
          IF (DUM.GE.AAMAX) THEN
```

```
            IMAX=I
            AAMAX=DUM
          END IF
16      CONTINUE
        IF (J.NE.IMAX)THEN
          DO 17 K=1,N
            EX=A(IMAX,K)
            A(IMAX,K)=A(J,K)
            A(J,K)=EX
17        CONTINUE
          D=-D
          VV(IMAX)=VV(J)
        END IF
        INDX(J)=IMAX
        IF(J.NE.N)THEN
        IF(A(J,J).EQ.(0.0D0,0.0D0)) PAUSE 'SINGULAR MATRIX.'
          EX=(1.0D0,0.0D0)/A(J,J)
          DO 18 I=J+1,N
            A(I,J)=A(I,J)*EX
18        CONTINUE
        END IF
19    CONTINUE
      IF(A(N,N).EQ.(0.0D0,0.0D0)) PAUSE 'SINGULAR MATRIX.'
      RETURN
      END
```

The next subroutine is a modification for complex variables of the routine LUBKSB in Press et al. (1986, p. 36) that uses the results of the LU decomposition and forward and backward substitution to solve a system of complex linear equations. The modification also allows the system to be solved for M right hand sides. It requires the output from CLU as the matrix A and the array INDX. B is a matrix, the columns of which are the M right hand side vectors.

```
      SUBROUTINE CLUBKS(A,N,NP,INDX,B,M)
      DIMENSION INDX(N)
      COMPLEX*16 A(NP,NP),SUM,B(NP,M)
      DO 15 K=1,M
        II=0
        DO 12 I=1,N
          LL=INDX(I)
          SUM=B(LL,K)
          B(LL,K)=B(I,K)
          IF (II.NE.0)THEN
```

```
          DO 11 J=II,I-1
            SUM=SUM-A(I,J)*B(J,K)
11        CONTINUE
        ELSE IF (SUM.NE.(0.0D0,0.0D0)) THEN
          II=I
        END IF
        B(I,K)=SUM
12    CONTINUE
      DO 14 I=N,1,-1
        SUM=B(I,K)
        IF(I.LT.N)THEN
          DO 13 J=I+1,N
            SUM=SUM-A(I,J)*B(J,K)
13        CONTINUE
        END IF
        B(I,K)=SUM/A(I,I)
14    CONTINUE
15  CONTINUE
    RETURN
    END
```

The B matrix is constructed from the transpose of the upper triangular factor of the unrotated initial state covariance matrix, output from subroutine INIT (page 200) as the array RCOV, and augmented by the vector g which is a vector of zeros with a one in the last position.

```
      DO 20 I=1,NAR
         DO 10 J=I,NAR
            B(I,J)=0.0D0
            B(J,I)=RCOV(I,J)
   10    CONTINUE
         B(I,NAR+1)=0.0D0
         IF(I.EQ.NAR)B(I,NAR+1)=1.0D0
   20 CONTINUE
```

Subroutine CLUBKS is now called with M=NAR+1. What is returned in the matrix B is

$$\left[\ \mathbf{C}^{-1}\mathbf{T}' \quad \boldsymbol{\kappa}\ \right],$$

where $\boldsymbol{\kappa}$ is in equation (6.19) and is needed for the covariance matrix of the random input to the state in equation (6.20). Notice that while the input matrix B is real, the output is complex, so the matrix B needs to be declared COMPLEX.

The final step in computing the initial state covariance matrix of the rotated state vector is to multiply the first NAR by NAR block of the matrix B that is output from the above subroutine to form

$$\tilde{\mathbf{P}}_c(t_1|0) = [\mathbf{C}^{-1}\mathbf{T}'] \, [\mathbf{C}^{-1}\mathbf{T}']',$$

since

$$\mathbf{P}(t_1|0) = \mathbf{T}'\mathbf{T}.$$

This multiplication involving complex arrays is

```
      DO 30 I=1,NAR
         DO 20 J=I,NAR
            COV(I,J)=(0.0D0,0.0D0)
            DO 10 K=1,NAR
               COV(I,J)=COV(I,J)+B(I,K)*DCONJG(B(J,K)
10          CONTINUE
            IF(I.EQ.J)THEN
               COV(I,J)=DREAL(C(I,J))
            ELSE
               COV(J,I)=DCONJG(COV(I,J))
            END IF
20       CONTINUE
30    CONTINUE
```

A.4 A multivariate subroutine

Listed below is a pilot version of a multivariate subroutine. This program undoubtedly has some errors, but a programmer can get a feeling for the complex arithmetic involved in fitting multivariate models.

```
      SUBROUTINE MCAR1(NP,P,LIKE)
C
C  MULTIVATIATE CONTINUOUS TIME FIRST ORDER AUTOREGRESSION
C
      PARAMETER(NR=13,NLPMAX=14,NLPM1=NLPMAX+1,NSMAX=789)
      PARAMETER(NMAX=1578,NDMAX=3,NDMAX1=NDMAX+1)
      PARAMETER(NDUMAX=6,NDMAX2=NDMAX*2,NVMAX=10)
      REAL P(NP),DATA(NMAX,NVMAX),MISS,LARGE(NDMAX)
      DOUBLE PRECISION LIKE,SSE,Z(NDMAX,NDUMAX),
     +  RR(NDMAX,NDMAX),B(NDMAX,NDMAX),A(NDMAX,NDMAX),
     +  INNOV(NDMAX2,NLPM1),XX(NLPM1,NLPM1),U(NDMAX,NDMAX),
     +  X(NDMAX,NLPM1),R(NDMAX,NDMAX),DT,
     +  COV(NDMAX,NDMAX+NLPM1)
```

```
      COMPLEX*16 E(NDMAX),C(NDMAX,NDMAX),TC(NDMAX,NDMAX),
     + G(NDMAX,NDMAX1),Q(NDMAX,NDMAX),TCOV(NDMAX2,NDMAX2),
     + HCOV(NDMAX,NDMAX2),EX(NDMAX),H(NDMAX,NDMAX2),EXX,
     + STATE(NDMAX2,NLPM1)
      INTEGER NS,NDEP(NDMAX),NTIME,NI(NDMAX),NIN(NDMAX),ND,
     + NIND(NDMAX,NVMAX,2),NOBS(NSMAX),NQ(NDMAX),IND(NR,4),
     + NINTER(NDMAX,NVMAX,3),IMISS(NDMAX),IPOLY(NDMAX)
      COMMON /TRANS/NQ,NS,DATA,ND,NDEP,NTIME,IPOLY,NI,NIND,
     + IND,XX,SSE,NOBS,NTOT,NIN,NINTER,MISS,LARGE,AVE,SCALE,
     + NLP
C
C  THE THREE ARGUMENT FORMAT IS SPECIFIED BY THE NONLINEAR
C  OPTIMIZATION PROGRAM
C
C    NP   - NUMBER OF PARAMETERS TO BE ESTIMATED
C    P    - ARRAY OF PARAMETER VALUES FOR WHICH -2 LN
C           LIKELIHOOD IS TO BE CALCULATED.  ARRAY MUST
C           BE INITIALIZED WITH INITIAL GUESSES AT
C           PARAMETERS
C    LIKE - VALUE OF -2 LN LIKELIHOOD CALCULATED BY
C           SUBROUTINE
C
C  THE PARAMETER STATEMENT DEFINES MAXIMUM ARRAY SIZES
C
C   NR     - MAXIMUM NUMBER OF NONLINEAR VARIANCE AND
C            COVARIANCE PARAMETERS
C   NLPMAX - MAXIMUM NUMBER OF LINEAR PARAMETERS INCLUDING
C            THE CONSTANT TERM FOR EACH DEPENDENT VARIABLE
C   NSMAX  - MAXIMUM NUMBER OF SUBJECTS
C   NMAX   - MAXIMUM NUMBER OF OBSERVATIONS ON ALL SUBJECTS
C   NDMAX  - MAXIMUM NUMBER OF DEPENDENT VARIABLES
C   NDUMAX - MAXIMUM NUMBER OF RANDOM EFFECTS
C   NVMAX  - MAXIMUM NUMBER OF VARIABLES
C
C  THE REMAINDER OF THE INFORMATION NEEDED TO CALCULATE
C  -2 LN LIKELIHOOD IS PASSED THROUGH COMMON.  THE ORDER
C  AND DIMENSION OF EVERYTHING IN COMMON MUST AGREE WITH THE
C  CALLING PROGRAM.
C
C   NQ    - ARRAY OF NUMBER OF RANDOM EFFECTS FOR EACH
C           DEPENDENT VARIABLE
C   NS    - NUMBER OF SUBJECTS
C   DATA  - TWO DIMENSIONAL ARRAY OF OBSERVATIONS
C   ND    - NUMBER OF DEPENDENT VARIABLES
```

```
C   NDEP  - ARRAY OF THE DEPENDENT VARIABLES' NUMBERS
C   NTIME - THE INDEPENDENT VARIABLE SUCH AS TIME
C   IPOLY - ARRAY OF ORDER OF POLYNOMIAL IN TIME FOR EACH
C           DEPENDENT VARIABLE
C   NI    - THE NUMBER IN COVARIABLES WHICH MAY OR MAY
C           NOT INCLUDE THE INDEPENDENT VARIABLE
C   NIND  - ARRAY OF THE COVARIABLES' NUMBERS AND THE ORDER
C           OF THE INTERACTION WITH TIME
C   IND   - NP BY 4 ARRAY OF INTEGERS INDICATING WHICH
C             PARAMETERS ARE TO BE ESTIMATED BY NONLINEAR
C             OPTIMIZATION,
C
C          IND(I,1)=1, PARAMETER I IS U(IND(I,2),IND(I,3))
C                                  FOR IND(I,2).LE.IND(I,3)
C          IND(I,1)=2, PARAMETER I IS A(IND(I,2),IND(I,3))
C          IND(I,1)=3, PARAMETER I IS G(IND(I,2),IND(I,3))
C                                  FOR IND(I,2).GE.IND(I,3)
C          IND(I,1)=4, PARAMETER I IS RR(IND(I,2),IND(I,3))
C                                  FOR IND(I,2).LE.IND(I,3)
C
C  XX     - NORMAL EQUATION MATRIX RETURNED TO THE MAIN
C           PROGRAM IN FACTORED FORM
C  SSE    - SUM OF SQUARES OF RESIDUALS RETURNED TO THE
C           MAIN PROGRAM.  SINCE NO VARIANCES ARE
C           CONCENTRATED OUT OF THE LIKELIHOOD, THIS SHOULD
C           BE DISTRIBUTED AS CHI-SQUARE WITH NTOT DOF
C  NTOT   - TOTAL NUMBER OF OBSERVATIONS AVAILABLE ON ALL
C           DEPENDENT VARIABLES
C  NIN    - ARRAY OF INDEPENDENT VARIABLES NUMBERS FOR EACH
C           DEPENDENT VARIABLE
C  NINTER - ARRAY OF INDEPENDENT VARIABLES NUMBERS FOR
C           INTERACTIONS FOR EACH DEPENDENT VARIABLE, AND
C           THE ORDER OF THE INTERACTION OF THIS COMBINED
C           VARIABLE WITH TIME
C  MISS   - MISSING VALUE CODE
C  LARGE  - ARRAY OF 'LARGE' STANDARD DEVIATIONS FOR EACH
C           DEPENDENT VARIABLE, USED TO INITIALIZE STATE
C           COVARIANCE MATRIX IN NONSTATIONARY SITUATIONS
C  AVE    - SETS ORIGIN OF INDEPENDENT VARIABLE.  USUALLY
C           THIS VARIABLE IS 0.0
C  SCALE  - SETS SCALE OF INDEPENDENT VARIABLES.  USUALLY
C           THIS IS 1.0
C  NLP    - NUMBER OF LINEAR FIXED EFFECTS FOR ALL
C           DEPENDENT VARIABLES
```

```
C
C INITIALIZE THE A, G, ROOT R AND U MATRICES, NDU IS THE
C DIMENSION OF THE U MATRIX
C
      WRITE(*,'(2H P,7G11.4/(2X,7G11.4))')(P(I),I=1,NP)
      NDU=0
      DO 40 I=1,ND
         NDU=NDU+NQ(I)
         DO 30 J=1,ND
            A(I,J)=0.0D0
            G(I,J)=0.0D0
            RR(I,J)=0.0D0
   30    CONTINUE
   40 CONTINUE
      IF(NDU.GT.0)THEN
         DO 44 I=1,NDU
            DO 42 J=1,NDU
               U(I,J)=0.0D0
   42       CONTINUE
   44    CONTINUE
      END IF
C
C ENTER CURRENT VALUES OF NONLINEAR PARAMETERS
C
      DO 46 I=1,NP
         IF(IND(I,1).EQ.1)U(IND(I,2),IND(I,3))=P(I)
         IF(IND(I,1).EQ.2)A(IND(I,2),IND(I,3))=P(I)
         IF(IND(I,1).EQ.3)G(IND(I,2),IND(I,3))=P(I)
         IF(IND(I,1).EQ.4)RR(IND(I,2),IND(I,3))=P(I)
   46 CONTINUE
C
C DIAGONALIZE THE A MATRIX USING AN IMSL SUBROUTINE FOR
C NONSYMMETRIC MATRICES
C
      CALL EIGRF(A,ND,NDMAX,1,E,C,NDMAX,TCOV,IER)
      IF(IER.NE.0)THEN
         WRITE(2,*)'IER=',IER,' IN EIGRF'
         STOP
      END IF
C
C MOVE EIGENVECTORS INTO TC AS TEMPORARY STORAGE SINCE
C SUBROUTINE LEQT1C DESTROYS THE INPUT MATRIX
C
      DO 49 I=1,ND
```

```
         DO 48 J=1,ND
            TC(I,J)=C(I,J)
   48    CONTINUE
   49 CONTINUE
C
C SOLVE SYSTEM OF COMPLEX EQUATIONS USING IMSL SUBROUTINE TO
C CALCULATE C**(-1)G AND C**(-1)LARGE
C
      DO 50 I=1,ND
         G(I,ND+1)=LARGE(I)
   50 CONTINUE
      CALL LEQT1C(TC,ND,NDMAX,G,ND+1,NDMAX,0,TCOV,IER)
      IF(IER.NE.0)THEN
         WRITE(2,*)'IER=',IER,' IN LEQT1C'
         STOP
      END IF
C
C CALCULATE (C**(-1)G)(C**(-1)G)'
C
      DO 86 I=1,ND
         DO 84 J=I,ND
            Q(I,J)=0.0D0
            DO 82 K=1,ND
               Q(I,J)=Q(I,J)+G(I,K)*DCONJG(G(J,K))
   82       CONTINUE
   84    CONTINUE
   86 CONTINUE
C
C CALCULATE THE BETWEEN SUBJECT COVARIANCE MATRIX
C
      IF(NDU.GT.0)THEN
         DO 56 J=1,NDU
            DO 54 I=1,J
               B(I,J)=0.0D0
               DO 52 K=1,I
                  B(I,J)=B(I,J)+U(K,I)*U(K,J)
   52          CONTINUE
               B(J,I)=B(I,J)
   54       CONTINUE
   56    CONTINUE
      END IF
C
C CALCULATE THE OBSERVATIONAL ERROR COVARIANCE MATRIX
C
```

```
      DO 66 J=1,ND
         DO 64 I=1,J
            R(I,J)=0.0D0
            DO 62 K=1,I
               R(I,J)=R(I,J)+RR(K,I)*RR(K,J)
   62       CONTINUE
            R(J,I)=R(I,J)
   64    CONTINUE
   66 CONTINUE
      NLP1=NLP+1
C
C CALCULATE THE DIMENSION OF THE STATE VECTOR
C
      NSTATE=ND+NDU
C
C INITIALIZE LIKE AND [X'X:X'Y] MATRIX
C
      LIKE=0.0D0
      DO 74 I=1,NLP1
         DO 72 J=I,NLP1
            XX(I,J)=0.0D0
   72    CONTINUE
   74 CONTINUE
C
C START SUBJECT LOOP
C
      NT=0
      NTOT=0
      DO 999 II=1,NS
C
C INITIALIZE TRANSFORMED STATE AND STATE COVARIANCE FOR
C SUBJECT II
C
         DO 78 I=1,NSTATE
            DO 76 J=1,NLP1
               STATE(I,J)=(0.0D0,0.0D0)
   76       CONTINUE
            DO 75 J=1,NSTATE
               TCOV(I,J)=(0.0D0,0.0D0)
   75       CONTINUE
   78    CONTINUE
         DO 94 I=1,ND
            DO 92 J=I,ND
               IF(DREAL(E(I)+DCONJG(E(J))).LT.0.0D0)THEN
```

```
                TCOV(I,J)=-Q(I,J)/(E(I)+DCONJG(E(J)))
             ELSE
                Q(I,J)=G(I,ND+1)*G(J,ND+1)
             END IF
   92      CONTINUE
   94   CONTINUE
        IF(NDU.GT.0)THEN
           DO 98 I=1,NDU
              DO 96 J=1,NDU
                 TCOV(ND+I,ND+J)=B(I,J)
   96         CONTINUE
   98      CONTINUE
        END IF
        DT=0.0
C
C  START WITHIN SUBJECT LOOP
C
        DO 250 IT=1,NOBS(II)
           NT=NT+1
           IF(IT.GT.1)DT=DATA(NT,NTIME)-DATA(NT-1,NTIME)
           IF(DT.LT.0.0)THEN
              WRITE(2,*)'TIME STEP NEGATIVE, SUBJECT',II
              WRITE(2,*)'OBSERVATION',IT
              STOP
           END IF
C
C SET UP X AND Z ARRAYS
C
           ICOL=0
           IZCOL=0
           DO 270 I=1,ND
              IMISS(I)=0
              DO 260 J=1,NLP
                 X(I,J)=0.0D0
  260         CONTINUE
              IF(NDU.GT.0)THEN
                 DO 261 J=1,NDU
                    Z(I,J)=0.0D0
  261            CONTINUE
              END IF
              ICOL=ICOL+1
              X(I,ICOL)=1.0D0
              IF(NQ(I).GT.0)THEN
                 IZCOL=IZCOL+1
```

```
                 Z(I,IZCOL)=1.0D0
               END IF
               IF(IPOLY(I).GT.0)THEN
                 DO 265 J=1,IPOLY(I)
                   STIME=(DATA(NT,NTIME)-AVE)*SCALE
                   ICOL=ICOL+1
                   X(I,ICOL)=X(I,ICOL-1)*STIME
                   IF(NQ(I).GT.J)THEN
                     IZCOL=IZCOL+1
                     Z(I,IZCOL)=X(I,ICOL)
                   END IF
 265             CONTINUE
               END IF
               IF(NI(I).GT.0)THEN
                 DO 280 J=1,NI(I)
                   IF(DATA(NT,NIND(I,J,1)).EQ.MISS)THEN
                     IMISS(I)=1
                     GO TO 270
                   END IF
                   ICOL=ICOL+1
                   X(I,ICOL)=DATA(NT,NIND(I,J,1))
                   IF(NIND(I,J,2).GT.0)THEN
                     DO 278 K=1,NIND(I,J,2)
                       ICOL=ICOL+1
                       X(I,ICOL)=X(I,ICOL-1)*STIME
 278                 CONTINUE
                   END IF
 280             CONTINUE
               END IF
               IF(NIN(I).GT.0)THEN
                DO 282 J=1,NIN(I)
                  IF(DATA(NT,NINTER(I,J,1)).EQ.MISS .OR.
     +              DATA(NT,NINTER(I,J,2)).EQ.MISS)THEN
                    IMISS(I)=1
                    GO TO 270
                  END IF
                  ICOL=ICOL+1
                  X(I,ICOL)=DATA(NT,NINTER(I,J,1))
     +                      *DATA(NT,NINTER(I,J,2))
                  IF(NINTER(I,J,3).GT.0)THEN
                    DO 281 K=1,NINTER(I,J,3)
                      ICOL=ICOL+1
                      X(I,ICOL)=X(I,ICOL-1)*STIME
 281                CONTINUE
```

```
                 END IF
 282            CONTINUE
             END IF
             IF(DATA(NT,NDEP(I)).EQ.MISS)THEN
                IMISS(I)=1
                GO TO 270
             END IF
             X(I,NLP1)=DATA(NT,NDEP(I))
 270      CONTINUE
C
C  STORE ARRAY OF EXPONENTIALS
C
          DO 80 I=1,ND
             EX(I)=CDEXP(DT*E(I))
C
C  PREDICT TRANSFORMED STATE
C
             DO 79 J=1,NLP1
                STATE(I,J)=EX(I)*STATE(I,J)
  79         CONTINUE
  80      CONTINUE
C
C  UPDATE TRANSFORMED STATE COVARIANCE MATRIX
C
          DO 100 I=1,ND
             DO 90 J=I,NSTATE
                TCOV(I,J)=TCOV(I,J)*EX(I)
                IF(J.LE.ND)THEN
                   TCOV(I,J)=TCOV(I,J)*DCONJG(EX(J))
                   EXX=E(I)+DCONJG(E(J))
                   IF(EXX.NE.(0.0D0,0.0D0))THEN
                      TCOV(I,J)=TCOV(I,J)+Q(I,J)*((EX(I)*
     +                DCONJG(EX(J))-(1.0D0,0.0D0)))/EXX
                   ELSE
                      TCOV(I,J)=TCOV(I,J)+G(I,J)*DT
                   END IF
                END IF
                IF(I.EQ.J)THEN
                   TCOV(I,J)=DREAL(TCOV(I,J))
                ELSE
                   TCOV(J,I)=DCONJG(TCOV(I,J))
                END IF
  90         CONTINUE
 100      CONTINUE
```

```
C
C  PREDICT OBSERVATIONS CHECKING FOR MISSING OBSERVATIONS
C  AND STORE RESIDUALS (INNOVATIONS) IN INNOV
C
            IC=0
            DO 160 I=1,ND
               IF(IMISS(I).EQ.1)GO TO 160
               NTOT=NTOT+1
               IC=IC+1
C
C FORM THE OBSERVATION MATRIX H
C
               DO 106 J=1,ND
                  H(IC,J)=C(I,J)
  106          CONTINUE
               IF(NDU.GT.0)THEN
                  DO 107 J=1,NDU
                     H(IC,ND+J)=Z(I,J)
  107             CONTINUE
               END IF
               DO 105 J=1,NLP1
                 INNOV(IC,J)=X(I,J)
                 DO 110 K=1,NSTATE
                  INNOV(IC,J)=INNOV(IC,J)-H(IC,K)*STATE(K,J)
  110            CONTINUE
  105          CONTINUE
C
C  CALCULATE H*TCOV AND STORE IN HCOV
C
               DO 130 J=1,NSTATE
                  HCOV(IC,J)=0.0D0
                  DO 120 K=1,NSTATE
                    HCOV(IC,J)=HCOV(IC,J)+H(IC,K)*TCOV(K,J)
  120             CONTINUE
  130          CONTINUE
C
C  CALCULATE INNOVATION COVARIANCE MATRIX
C
               JC=0
               DO 150 J=1,I
                  IF(IMISS(J).EQ.1)GO TO 150
                  JC=JC+1
                  COV(IC,JC)=0.0D0
                  DO 140 K=1,NSTATE
```

```
                  COV(IC,JC)=COV(IC,JC)+HCOV(IC,K)
     +                              *DCONJG(H(JC,K))
  140               CONTINUE
                    COV(IC,JC)=COV(IC,JC)+R(I,J)
                    COV(JC,IC)=COV(IC,JC)
  150             CONTINUE
  160         CONTINUE
              IF(IC.EQ.0)GO TO 250
C
C  FACTOR INNOVATION COVARIANCE MATRIX AUGMENTED BY
C  INNOVATIONS
C
              DO 165 I=1,IC
                 DO 164 J=1,NLP1
                    COV(I,IC+J)=INNOV(I,J)
  164            CONTINUE
  165         CONTINUE
              CALL FACTOR(COV,IC,NDMAX,IC+NLP1,IER)
              IF(IER.NE.0)THEN
                 WRITE(2,3000)
 3000            FORMAT(/' MATRIX NOT POSITIVE DEFINITE IN
     + FACTOR'/' CALLED BY MCAR')
                 STOP
              END IF
C
C  CALCULATE DETERMINANT PART OF -2 LN LIKELIHOOD
C
              DO 170 I=1,IC
                 LIKE=LIKE+DLOG(COV(I,I)**2)
  170         CONTINUE
C
C ACCUMULATE NORMAL EQUATIONS
C
              DO 174 I=1,NLP1
                 DO 172 J=I,NLP1
                    DO 171 K=1,IC
                     XX(I,J)=XX(I,J)+COV(K,IC+I)*COV(K,IC+J)
  171               CONTINUE
  172            CONTINUE
  174         CONTINUE
C
C  CALCULATE T' INVERSE * HCOV BY FORWARD ELIMINATION
C
              DO 200 J=1,NSTATE
```

```
                HCOV(1,J)=HCOV(1,J)/COV(1,1)
                IF(NSTATE.LE.1)GO TO 200
                DO 190 I=2,ND
                   EXX=(0.0D0,0.0D0)
                   DO 180 K=1,I-1
                      EXX=EXX+COV(K,I)*HCOV(K,J)
  180              CONTINUE
                   HCOV(I,J)=(HCOV(I,J)-EXX)/COV(I,I)
  190           CONTINUE
  200        CONTINUE
C
C  UPDATE TRANSFORMED STATE AND COVARIANCE MATRIX
C
             DO 240 I=1,NSTATE
                DO 205 J=1,NLP1
                   DO 210 K=1,IC
                     STATE(I,J)=STATE(I,J)+DCONJG(HCOV(K,I))
     +                                       *COV(K,IC+J)
  210              CONTINUE
  205           CONTINUE
                DO 230 J=I,NSTATE
                   DO 220 K=1,IC
                      TCOV(I,J)=TCOV(I,J)-DCONJG(HCOV(K,I))
     +                                      *HCOV(K,J)
  220              CONTINUE
                   IF(I.EQ.J)THEN
                      TCOV(I,J)=REAL(TCOV(I,J))
                   ELSE
                      TCOV(J,I)=DCONJG(TCOV(I,J))
                   END IF
  230           CONTINUE
  240        CONTINUE
  250     CONTINUE
  999 CONTINUE
C
C FACTOR NORMAL EQUATIONS TO OBTAIN RESIDUAL SUM OF SQUARES
C
      SSE=XX(NLP1,NLP1)
      CALL FACTOR(XX,NLP,NLPM1,NLP1,IER)
      IF(IER.NE.0)THEN
         WRITE(2,3001)
 3001    FORMAT(/' MATRIX NOT POSITIVE DEFINITE IN FACTOR'/
     *          ' FACTORING XX')
         STOP
```

```
      END IF
      DO 760 I=1,NLP
         SSE=SSE-XX(I,NLP1)**2
  760 CONTINUE
C
C CALCULATE -2 LN LIKELIHOOD
C
      LIKE=LIKE+SSE+NTOT*1.837877067D0
      WRITE(*,'(7H LIKE= ,F10.2)')LIKE
      RETURN
      END
```

References

Akaike, H. (1973a) Information theory and an extension of the maximum likelihood principle. *Second International Symposium on Information Theory*, (B. N. Petrov and F. Csaki, Eds.), Budapest: Akademia Kaido, 267–81.

Akaike, H. (1973b) Maximum likelihood identification of Bayesian autoregressive moving average models. *Biometrika*, **60**, 255–65.

Akaike, H. (1974a) A new look at the statistical model identification. *IEEE Trans. on Automatic Control*, **AC-19**, 716–23.

Akaike, H. (1974b) Markovian representation of stochastic processes and its applications to the analysis of autoregressive moving average processes. *Ann. Inst. Statist. Math.*, **26**, 363–87.

Akaike, H. (1975) Markovian representation of stochastic processes by cononical variables. *SIAM J. Control*, **13**, 162–73.

Ansley, C.F. and Kohn, R. (1985) Estimation, filtering and smoothing in state space models with incompletely specified initial conditions. *Ann. Statist.*, **13**, 1286–316.

Aoki, M. (1987) *State Space Modeling of Time Series*, Springer-Verlag, Berlin.

Bierman, G.J. (1977) *Factorization Methods for discrete sequential estimation*, Series: Mathematics in Science and Engineering, Vol. 128, Academic Press, New York.

Boudjellaba, H., Dufour, J.-M. and Roy, R. (1992) Testing causality between two vectors in multivariate autoregressive moving average models. *J. Am. Statist. Assoc.*, **87**, 1082–90.

Box, G.E.P. and Jenkins, G.M. (1976) *Time Series Analysis, forecasting and control*, Revised Edition, Holden-Day, San Francisco.

Box, G. E. P. and Tiao, G. C. (1973) *Bayesian Inference in Statistical Analysis*, Addison-Wesley, Reading, Massachusetts.

Bozdogan, H. (1987) Model selection and Akaike's information Criterion (AIC): the general theory and its analytical extensions. *Psychometrika*, **52**, 345–70.

Chi, E.M. and Reinsel, G.C. (1989) Models for longitudinal data with random effects and AR(1) errors. *J. Am. Statist. Assoc.*, **84**, 452–9.

Crowder, M.J. and Hand, D.J. (1990) *Analysis of Repeated Measures*, Chapman and Hall, London.

De Jong, P. (1988) The likelihood for a state space model. *Biometrika*, **75**, 165–9.

De Jong, P. (1991a) Stable algorithms for the state space model. *J. Time Series Analysis*, **12**, 143–57.

De Jong, P. (1991b) The diffuse Kalman filter. *Ann. Statist.*, **19**, 1073–83.

Dennis, J.E. and Schnabel, R.B. (1983) *Methods for Unconstrained Optimization and Nonlinear Equations*, Prentice-Hall, Englewood Cliffs, N. J.

Diggle, P.J. (1988) An approach to the analysis of repeated measurements. *Biometrics*, **44**, 959–71.

Diggle, P.J. (1989) Testing for random dropouts in repeated measurement data. *Biometrics*, **45**, 1255–8.

Diggle, P.J. (1990) *Time Series: A Biostatistical Introduction*, Oxford University Press, Oxford.

Doob, J.L. (1953) *Stochastic Processes*, Wiley, New York.

Draper, N.R. and Smith, H. (1981) *Applied Regression Analysis*, Second Edition, Wiley, New York.

Duncan, D.B. and Horn, S.D. (1972) Linear dynamic recursive estimation from the viewpoint of regression analysis. *J. Am. Statist. Assoc.*, **67**, 815–21.

Duong, Q.P. (1984) On the choice of the order of autoregressive models: a ranking and selection approach. *J. Time Ser. Analysis*, **5**, 145–57.

Durbin, J. (1960) The fitting of time series models. *Int. Statist. Rev.*, **28**, 233–44.

Gabow, P. A., Johnson, A. M., Kaehny, W. D., Kimberling, W. J, Lezotte, D. C., Duley, I. T. and Jones, R. H. (1992) Factors affecting the progression of renal disease in autosomal-dominant polycystic kidney disease. *Kidney International*, **41**, 1311-9.

Gabriel, K.R. (1962) Ante-dependence analysis of an ordered set of variables. *Ann. Math. Statist.*, **33**, 201–12.

Golub, G. H. and Van Loan, C. F. (1983) *Matrix Computations*, The Johns Hopkins University Press, Baltimore.

Graybill, F.A. (1976) *Theory and Application of the Linear Model*, Duxbury Press, North Scituate, Massachusetts.

Grizzle, J.E. and Allen, D.M. (1969) Analysis of growth and dose response curves. *Biometrics*, **25**, 357–82.

Hamman, R.F., Marshall, J.A., Baxter, J., Kahn, L.B., Mayer, E.J., Orleans, M., Murphy, J.R. and Lezotte, D.C. (1989) Methods and prevalence of non-insulin-dependent diabetes mellitus in a biethnic Colorado population: The San Luis Valley Diabetes Study. *Am. J. Epidemiol.*, **129**, 295–311.

Harvey, A.C. (1981) *Time Series Models*, Philip Allan, Oxford.

Harvey, A.C. and Phillips, G.D.A. (1979) Maximum likelihood estimation of regression models with autoregressive-moving average disturbances. *Biometrika*, **66**, 49–58.

Harville, D.A. (1974) Bayesian inference for variance components using only error contrasts. *Biometrika*, **61**, 383–5.

Harville, D.A. (1976) Extensions of the Gauss-Markov theorem to include the estimation of random effects. *Ann. Statist.*, **4**, 384–95.

Harville, D.A. (1977) Maximum likelihood approaches to variance component estimation and to related problems. *J. Am. Statist. Assoc.*, **72**, 320–40.

Herbach, L.H. (1959) Properties of model II—type analysis of variance tests. *Ann. Math. Statist.*, **30**, 939–59.

Hirst, K., Zerbe, G.O., Boyle, D.W. and Wilkening, R.B. (1991) On nonlinear random effects models for repeated measurements. *Commun. Statist.*, **B20**, 463–78.

Hocking, R.R. (1985) *The Analysis of Linear Models*, Brooks/Cole Publishing Company, Monterey, California.

Jennrich R.I. and Schluchter, M.D. (1986) Unbalanced repeated-measures models with structured covariance matrices. *Biometrics*, **42**, 805–20.

Jones, R.H. (1964) Predicting multivariate time series. *Journal of Applied Meteorology* **3**, 285–9.

Jones, R.H. (1980) Maximum likelihood fitting of ARMA models to time series with missing observations. *Technometrics*, **22**, 389–95.

Jones, R.H. (1981) Fitting continuous time autoregressions to discrete data. *Applied Time Series Analysis II*, (D. F. Findley, editor), Academic Press, 651–82.

Jones, R.H. (1984) Fitting multivariate models to unequally spaced data. *Time Series Analysis of Irregularly Observed Data* (E. Parzen, Ed.), *Lecture Notes in Statistics*, **25**, Springer-Verlag, 158–88.

Jones, R.H. (1985a) Repeated measures, interventions and time series analysis. *Psychoneuroendocrinology*, **10**, 5–14.

Jones, R. H. (1985b) Time series analysis with unequally spaced data. *Handbook of Statistics, Vol. 5: Time Series in the Time Domain* (E. J. Hannan, P. R. Krishnaiah and M. M. Rao, Eds.), North-Holland, 157-177.

Jones, R.H. (1986) Time series regression with unequally spaced data. *Essays in Time Series and Allied Processes, J. Appl. Prob.*, Special Volume 23A (Gani, J. and Priestley, M. B., Eds), The University, Sheffield S3 7RH, England, 89–98.

Jones, R.H. (1990) Serial correlation or random subject effects? *Commun. Statist. Simula.*, **19**, 1105–23.

Jones, R. H. (1993) Longitudinal data models with fixed and random effects. *Proceedings of the First US/Japan Conference on the Fro- tiers of Statistical Modeling: An Informational Approach, Volume I: Theory and Methodology of Time Series Analysis*, Kluwer Academic Publishers, Dordrecht, Netherlands, in press.

Jones, R.H. and Ackerson, L.M. (1990) Serial correlation in unequally spaced longitudinal data. *Biometrika*, **77**, 721–31.

Jones, R.H. and Boadi-Boateng, F. (1991) Unequally spaced longitudinal data with AR(1) serial correlation. *Biometrics*, **47**, 161–75.

Jones, R.H., Daniels, A. and Bach, W. (1976) Fitting a circular distribution to a histogram. *J. Appl. Meteorology*, **15**, 94–8.

Jones, R.H. and Molitoris, B.A. (1984) A statistical method for determining the breakpoint of two lines. *Analytical Biochemistry*, **141**, 287-90.

Jones, R.H., Reeve, E.B. and Swanson, G.D. (1984) Statistical identification of compartmental models with applications to plasma protein kinetics. *Computers and Biomedical Research*, **17**, 277–88.

Jones, R.H. and Tryon, P.V. (1987) Continuous time series models for unequally spaced data applied to modeling atomic clocks. *SIAM Journal on Scientific and Statistical Computing* **8**, 71–81.

Kalman, R.E. (1960) A new approach to linear filtering and prediction problems. *Trans. ASME J. Basic Eng.*, **82D**, 35–45.

Kalman, R.E. and Bucy, R.S. (1961) New results in linear filtering and prediction theory. *Tran. ASME J. Basic Eng.*, **83D**, 95–108.

Kenward, M.G. (1987) A method for comparing profiles of repeated measurements. *Appl. Statist.*, **36**, 296–308.

Kitagawa, G. (1987) New developments in time series modelling and analysis (Discussion). *Bull. Int. Statist. Inst.*, Proceedings of the 46th Session, Tokyo, 8-16 September, 1987, Book 4, 137.

Laird, N.M. (1982) Computation of variance components using the EM algorithm. *J. Statist. Comput. Simul.*, **14**, 295–303.

Laird, N.M. (1988) Missing data in longitudinal studies. *Statist. Med.*, **7**, 305–15.

Laird, N.M. and Ware, J.H. (1982) Random effects models for longitudinal data *Biometrics*, **38**, 963–74.

Levinson, N. (1947) The Wiener RMS error criterion in filter design and prediction *J. Math. Physics*, **25**, 261–78.

Liang, K.Y. and Zeger, S.L. (1986) Longitudinal data analysis using generalized linear models. *Biometrika*, **73**, 13–22.

Lindstrom, M.J. and Bates, D.M. (1990) Nonlinear mixed effects models for repeated measures data. *Biometrics*, **46**, 673–87.

Little, R.J.A. and Rubin, D.B. (1987) *Statistical Analysis with Missing Data*. Wiley, New York.

Matthews, J.N.S., Altman, D.G., Campbell, M.J. and Royston, P. (1990) Analysis of Serial measurements in medical research. *Br. Med.*

J., **300**, 230–5.

McCullagh, P. and Nelder, J.A. (1989) *Generalized Linear Models.* (2nd edn). Chapman and Hall, London.

Moler C. and Van Loan, C. (1978) Nineteen dubious ways to compute the exponential of a matrix. *SIAM Review*, **20**, 801–36.

Nelder, J.A. (1987) An extended quasi-likelihood function. *Biometrika*, **74**, 221–32.

Pandit, S.M. and Wu, S-M. (1983) *Time Series and System Analysis with Applications*, Wiley, New York.

Potthoff, R.F. and Roy, S.N. (1964) A generalized multivariate analysis of variance model useful especially for growth curve problems. *Biometrika*, **51**, 313–26.

Prentice, A.M. (1990) The doubly-labelled water method: Technical recommendations for use in humans, a consensus report by the IDECG working group. Conference held September, 1988, Claire College, Cambridge.

Press, W.H., Flannery, B.P., Teukolsky, S.A. and Vetterling, W.T. (1986) *Numerical Recipes: The Art of Scientific Computing*, Cambridge University Press, Cambridge.

Rao, C.R. (1959) Some problems involving linear hypotheses in multivariate analysis. *Biometrika*, **46**, 49–58.

Rao, C.R. (1965) The theory of least squares when the parameters are stochastic and its application to the analysis of growth curves. *Biometrika*, **52**, 447–58.

Rao, C.R. (1973) *Linear Statistical Inference and Its Applications*, second edition, Wiley, New York.

Ridout, M.S. (1991) Reader reaction to: Testing for random dropouts in repeated measurement data, by P. Diggle. *Biometrics*, **47**, 1617–21.

Robinson, D.L. (1987) Estimation and use of variance components. *The Statistician*, **36**, 3-14.

Robinson, G.K. (1991) That BLUP is a good thing: the estimation of random effects. *Statistical Science*, **6**, 15–51.

Rubin, D.B. (1976) Inference and missing data. *Biometrika*, **63**, 581–92.

Rudemo, M., Ruppert, D. and Streibig, J.C. (1989) Random effects models in nonlinear regression with applications to bioassay. *Biometrics*, **45**, 349–62.

Ruppert, D., Cressie, N. and Carroll, R.J. (1989) A transformation/weighting model for estimating Michaelis-Menten parameters. *Biometrics*, **45**, 637–56.

Sakamoto, Y., Ishiguro, M. and Kitagawa, G. (1986) *Akaike Information Criterion Statistics*, Kluwer Academic Publishers, Dordrecht, Holland.

SAS (1985) *SAS/STAT Guide for Personal Computers, Version 6 Edi-*

tion, SAS Institute Inc., Cary, NC

Schwarz, G. (1978) Estimating the dimension of a model. *Ann. Statist.*, **6**, 461–4.

Schweppe, F.C. (1965) Evaluation of likelihood functions for Gaussian signals. *IEEE Trans. Inform. Theory*, **11**, 61–70.

Sheiner, L.B. and Beal, S.L. (1980) Evaluation of methods for estimating population pharmacokinetic parameters. I. Michaelis-Menten model: routine clinical pharmacokinetic data. *J. Pharmacokin. Biopharm.*, **8**, 553–71.

Srivastava, V.K. and Giles, D.E.A. *Seemingly Unrelated Regression Equations Models*, Marcell Dekker, New York.

Stein, C. (1956) Inadmissibility of the usual estimator for the mean of a multivariate normal mean. *Proc. Third Berkeley Symp.*, **1**, 197–206, Berkeley: University of California Press.

Stiratelli, R., Laird, N. and Ware, J.W. (1984) Random-effects models for serial observations with binary response. *Biometrics*, **40**, 961–71.

Thompson, W.A. (1962) The problem of negative estimates of variance components. *Ann. Math. Statist.*, **33**, 273–89.

Vonesh, E.F. and Carter, R.L. (1992) Mixed-effects nonlinear regression for unbalanced repeated measures. *Biometrics*, **48**, 1–17.

Watson, G.S. (1967) Linear least squares regression. *Ann. Math. Statist.*, **38**, 1679–99.

Wedderburn, R.W.M. (1974) Quasi-likelihood functions, generalized linear models and the Gauss-Newton method. *Biometrika*, **61**, 439–47.

Whittle, P. (1963) On the fitting of multivariate autoregressions and the approximate canonical factorization of a spectral density matrix. *Biometrika* **50**, 129–34.

Wiberg, D.M. (1971) *State Space and Linear Systems*, Schaum's Outline Series, McGraw-Hill, New York.

Winer, B.J. (1971) *Statistical Principles in Experimental Design*, second edition, McGraw-Hill, New York.

Zeger, S.L. and Liang, K.Y. (1986) Longitudinal data analysis for discrete and continuous outcomes. *Biometrics*, **42**, 121–30.

Zerbe, G.O. and Jones, R.H. (1980) On application of growth curve techniques to time series data. *J. Am. Statist. Assoc.*, **75**, 507–9.

Index